THE SMART HOME MANUAL

How To Automate Your Home To Keep Your Family ***Entertained***, ***Comfortable***, And ***Safe***

MARLON BUCHANAN

HomeTechHacker.com

ISBN: 978-1-7355430-0-0 (paperback)
ISBN: 978-1-7355430-1-7 (ebook)

Edited by Graham Southorn

Contents

PREFACE

I'VE WANTED TO MAKE TECHNOLOGY work for me since I was a young child. The advanced technology of TV shows like *The Jetsons* and *Star Trek* inspired my interest in gadgets that make home life more entertaining and fulfilling. As a child, the first home automation I ever created was programming the cable box to change the channel to ESPN at 8:00 p.m. every day so I could watch SportsCenter. Not everyone in the house was happy with that programming, but I thought it was great!

My passion for technology remained a constant force in my life. I've spent most of my career in technology—working as an IT help desk technician, a software developer, a business systems analyst, and eventually an IT manager. Although I love that IT management allows me to strategically direct the work of others, I really miss the hands-on technical work that I used to do.

I've filled that hole by taking on various technical projects around my house. I first started building my smart home around 2006. I spent a lot of time trying to stream media throughout my house. I ended up learning a lot about home networking, cybersecurity, and streaming media devices. I learned the value of automation and remote access to my home as I branched out into installing timers for my outdoor lights, a W-iFi-enabled thermostat, and an Internet-connected home security system.

My smart home projects exploded when we moved into a new home eight years ago. I imagined all sorts of smart home functionality I could add to make our lives more convenient. Ours being a new home, I wanted to do everything right. So I took the time to learn about smart home hubs, smart home protocols, and all the various smart home devices out there, and put together a plan. Piece by piece, I added smart lights, smart plugs, smart switches, smart door locks—the list goes on.

My family was a little hesitant about my projects at first. What if something goes wrong and our lights stop working? What if you hurt yourself installing a smart switch? Over time, these fears completely disappeared, and now my family has become reliant on our smart home technology and security. They love knowing that lights around the house will come on and turn off at the right times without any intervention. They love being able to turn on lights, TVs, the fireplace, and even the robot vacuum with a simple voice command. They feel safer knowing that we have a 24/7 surveillance system, doors that lock on their own in case we forget to lock them at night, and panic buttons placed around the house that silently send distress messages to loved ones.

I'm very proud that my family can use the features of our smart home without a manual. Everything just works, and I think they take a lot of it for granted now. I'm also proud that even when the Internet connection is down many things continue to work. The worst that happens is that some things operate as they would in a normal home.

I wrote this book because I want others to share in the joy, convenience, and security of a smart home. I want to help people get off to the best start possible in achieving their smart home goals. I hope this book provides you with education and inspiration, eliminating your frustration. Technology has advanced in ease of use to the point where anyone can build a smart home. Now is the time to start building the smart home of your dreams!

ABOUT THIS BOOK

Whom Is This Book for?

ALEXA. GOOGLE ASSISTANT. SMART SWITCHES. Smart plugs. Smart lights. Smart, smart, smart, smart! These days, it's hard to turn on the TV, browse the web, or consume social media without being inundated with ads and articles about smart devices for your home. There's a good reason why smart devices get all this attention. They're impacting our lives more and more every day.

The collection and collaboration of smart devices in your home make up what is commonly referred to as a smart home. Smart homes help people keep tabs on their families and homes. Smart homes help keep people safe, warm, and secure. Smart homes lift the burdens of mundane tasks that we get tired of or have a hard time remembering to do. Most importantly, smart homes make our lives better.

This book is for those of you who've gotten the smart home bug. You want the convenience and cool factor that you hear and see in the media, but you are unsure where to start. Or, you've got a few smart devices and you want to know how to make them work together so you can make the most of them. Perhaps you've been tinkering for years and want to take your home automation to the next level. If any of these sentences describe you, then this book is for you.

By the time you are done reading this book, you'll know the following:

- What a smart home is and what it can do for you
- How much a smart home costs
- How to get started building a smart home from scratch
- How to pick the right smart home devices for your situation
- How to plan for the future of the smart home
- How to secure your smart home (very important!)

While this book focuses on those who want to build their smart homes themselves, it is also useful for those who want to pay someone to do everything. Having knowledge of the smart home concepts, technologies, and capabilities will help you direct your smart home builder to create the best result for you. If any of this sounds interesting to you, keep reading!

How to Use This Book

This book steps you through building a smart home from the ground up. You can read through it sequentially and learn everything you need to go from zero to smart home in no time flat. This book has information that helps you build the foundation of your smart home, including planning, installing, and enhancing your smart home.

However, many of you will have some of the pieces put together already. Maybe you already have a home network you feel provides you with enough speed and security. Maybe you already have a few of the smart home pieces together and you're just looking for inspiration to make everything work better together or improve on your smart home capabilities. In that case, this book can also be used as a reference. You can skip right to the sections you want for everything you need.

Also, be sure to take advantage of the glossary of terms, checklists, and additional resources in the appendix. Smart home technology involves a lot of jargon, and the glossary is really helpful with explaining technical terms. The checklists give you a quick way to track accomplishing some of the recommendations in this book. Finally, the additional resources that the appendix provides are links to up-to-date smart home web resources. These sections come in handy even after you have read the entire book.

Conventions Used in This Book

Here are the conventions used in this book, which highlight important information:

- **What I Do**: In several sections of this book, I give specific information about how I address the topics discussed in the context of my own home. For example, in the home network section, I discuss the router I chose and some of the security steps I've taken in my own home. This is intended to give you a real-world, practical example of something you can do.

- **Key Takeaways**: At the end of each part, I provide a checklist of the key information that you can use as a summary for quick reference.
- **A Day in the Life:** Each part also ends with an example of what life could be like in your smart home.

Part 1

Introduction to Smart Homes

What Is a Smart Home?

PEOPLE DEFINE SMART HOMES IN different ways. To me, a smart home is a home that provides some combination of comfort, energy efficiency, security, lighting, etc., aided by technology that allows these systems to be automated, integrated, and available for remote control. Some examples of smart home functions may be turning on your outdoor lights automatically at night, being sent notifications when a visitor is at your door, and automating the house temperature to the right level when you're home.

When a lot of people first get into smart homes, they focus on remote control and remote monitoring. They want to control and monitor the temperature of their home when they are on vacation. Voice assistants like Alexa allow them to turn lights on and off and play music with simple commands. An app or two on their phone allows them to see who is at the door and remotely unlock it. Remote control is great, but it is really just another way to accomplish mundane tasks that you already do in a different way and from different locations.

Automation is what really makes a home smart. Instead of having to remember when to turn lights on, or to arm the alarm, or to make sure the doors are locked, you create automations that do all of this for you at the right times or under the right circumstances. Some examples of smart home automations include the following:

- Automatically arming your alarm, locking your doors, and turning off your lights when your home is empty
- Automatically turning on lights and sounds at certain times of the day to make it look like someone is home when you are away on vacation
- Having coffee made for you and playing your favorite new podcasts when you wake up in the morning
- Having a robot vacuum the floors when no one is home

As you read through this book, you'll see many more examples of automations you can do for your home. The point is that while remote control of your home is nice, automation is what makes a smart home truly worth the effort.

All kinds of smart home devices exist, and new devices come to market literally every single day. Typical smart home devices that people buy include the following:

- **Smart lighting** – These include smart bulbs like LIFX, Philips Hue, and Kasa (TP-Link). But these aren't just limited to bulbs. Some people buy or build LED light strips and strings to add accents around the house and to spice up holiday lighting displays.

- **Smart switches** – These, in some ways, are an alternative to smart lights. They allow you to remotely control and automate your existing lights by making the light switches smart. Lutron, Kasa, and Wemo are popular smart switch brands. Some smart switches can also be used with dimmable lighting.
- **Smart plugs** – Wemo and Kasa make popular smart plugs that allow for smart automation of outlets. Want to turn a lamp on remotely or at a certain time? Start a coffee maker? Monitor how much power a particular device uses? Smart plugs can do all of this and more.
- **Smart locks** – Smart locks automate door locks like deadbolts. Yale, Schlage, August, and Kwikset make popular models that allow you to open your door without a key and remotely check and lock your door for security.
- **Smart thermostats** – Nest and ecobee make really popular thermostats that automate your home's temperature. These devices can increase the comfort of your home at the right time and save you money on heating and cooling costs.
- **Voice assistants** – Voice assistants like Alexa, Siri, and Google Assistant have taken over our lives. Beyond answering trivia questions, telling jokes, and playing music, these devices can actually tie your other smart devices together in your home to create a great automated experience. In a sense, they can function as a smart home hub, and they may be the only smart home hub you need. You'll learn more about this later in the book.

One key thing to remember: Today's smart home is just the regular home of the future. In the not-too-distant future, people will take the features of today's smart home for granted, as they will be ubiquitous.

Why Would You Want a Smart Home?

It can take a lot of time to build a smart home. You should have some good reasons for putting in all the work (which could be useful in justifying time and costs to your significant other). Fortunately, there are a lot of good reasons to build a smart home. I'm going to detail a few of them for you.

Enhance and Enrich Your Life

A smart home should make your life easier by handling many of the mundane things you don't feel like doing or forget to do. For example, most people have morning and night routines they do over and over, so why not automate them? Build a smart home that slowly turns on the lights you need when you wake up. You could even have the lights start off very soft and then get brighter while playing music as your alarm clock. Have the coffee maker start, play the news, or turn on the TV.

When you get home from work, your house can know that you are home. When you arrive, it can welcome you with an update, turn on certain lights, turn up the thermostat (or turn on the A/C). A robot vacuum can clean your floors while you are gone. There is so much you can automate to make your life easier.

Safety and Security

Security is one of the best reasons to build a smart home. You can use security cameras to monitor the inside and outside of your home. Take it a step further. Use a smart doorbell (e.g., Nest Hello) to alert you when someone is at your home. Or when a package has been delivered.

Lighting is another great tool for securing your home. You can use smart lights outside, which automatically turn on when it gets dark and turn off at sunrise. Other security lights outside can be set to turn on when motion is detected. Set inside lights to come on and off according to schedules that mimic your being home when you are on vacation.

Do you ever forget to lock your doors at night or arm your security system? Your smart home can do it for you automatically. It can even make sure your garage doors are closed!

Don't forget about safety! Build a smart home that alerts you to water leaks, fires, and high carbon monoxide levels, using smart leak detectors and smart smoke and carbon monoxide detectors.

Save Money

Recouping the costs to build a smart home may quite possibly take years. Cost savings may not be the primary factor in building a smart home, but a smart home can help you save money. A smart thermostat (e.g., an ecobee or Nest) can make sure your heat and A/C are used in efficient ways by learning your patterns. Ceiling fans can be used smartly to distribute air for warming or cooling purposes.

Other cost savings can come from reducing power usage. Your smart home can turn lights off when you are no longer home or when you are going to sleep. You can also cut power to "vampire" devices that constantly suck energy, even when they aren't in use.

Voice Assistants

Voice assistants add a lot to a regular home. Having your calendar, the weather, and traffic at your fingertips is great. They also allow you to play games and listen to music. They function as intercoms and try to answer whatever you are curious about.

You significantly enhance the capabilities of your voice assistant when you add compatible smart home devices to your home. Combining a voice assistant with smart home devices allows you to control your TV, lights, security system, fans, thermostat, and so much more with just your voice.

Fun and Leisure

Building a smart home can be fun for those of us who lean towards the geeky side. I've learned a ton of technologies while building and improving my smart home. While at times building my smart home has been frustrating, overall, I am very happy with what my home can do. I also feel that keeping up with these technologies is important, as they continue to evolve and become a bigger part of our daily lives in and outside of the home.

Building a smart home can give you practice coding as well as learning the wants and needs of your family. Over time, you'll be proud of the automations and remote-control capabilities you've given your home.

Cool Factor

Sometimes you just want to show off to your friends and guests. Few things are cooler to show off than automated lights or being able to give voice commands to do common chores. A smart home can easily impress people!

You're Building a New Construction Home

If you are building a new home, don't miss out on a great opportunity to make your home smart. This is the easiest time to plan and install smart home devices. Even do-it-yourself (DIY) devices that you will eventually configure can be professionally installed by builders. As you'll see in the next section, the cost of building a smart home is a drop in the ocean compared to an overall home build price.

It Can Increase Your Home Resale Value

A 2015 survey by *Better Homes and Gardens* found that 64 percent of the millennials they surveyed were interested in having smart technology in their homes.[1] This generation

---

1 Cision PR Newswire, "Exclusive Survey from Better Homes and Gardens Reveals 'Next Gen' Homeowners Want Living Spaces That Combine 'Smart Homes and Smart Design'" provided by Better Homes and Gardens, January 21, 2015, accessed July 26, 2020, https://www.prnewswire.com/news-releases/exclusive-survey-from-better-homes-and-gardens-reveals-next-gen-homeowners-want-living-spaces-that-combine-smart-homes-and-smart-design-300023619.html.

represents the largest share of potential home buyers, and thus their desires will shape the market. A 2018 survey of potential buyers by Coldwell Banker found that 77 percent of those surveyed wanted smart thermostat pre-installed, 75 percent would want a smart fire detector pre-installed, and 63 percent would want a smart lighting system pre-installed.[2]

Some may be worried about the difficulty of learning and maintaining the smart devices in a new home. Others may have worries about selling a home with outdated smart home technology. These are legitimate concerns. You can mitigate these concerns by installing easy-to-use and easily replaceable devices.

If you do install smart features and sell your house, you'll want to be clear with potential buyers what's included and what's not. For example, they may see a smart thermostat in the home and will expect it to be there when they move in. If you are planning on taking it with you, that should be in writing. Any smart devices you leave behind should be reset back to factory settings and all your data deleted from them. This is for your privacy protection and allows the new owners to be able to set up the devices for themselves.

Smart home features are more likely to increase your home's value than decrease it. You may have to educate buyers. One real estate agent I know put demonstrations of a home's smart features on a loop on a TV as potential buyers visited the home. As I wrote earlier, the smart home

---

2 "Virtual Reality Poised to Revolutionize Home Buying, According to Coldwell Banker Real Estate Survey: Results of the 2018 Coldwell Banker Real Estate Smart Home Marketplace Survey," Coldwell Banker Real Estate LLC (blog), January 2018, accessed July 26, 2020, https://blog.coldwellbanker.com/wp-content/uploads/2018/01/Smart-Home-Survey-Data-Sheet-January-2018.pdf.

of today is just the "home" of tomorrow. Much of what we consider smart today will be what people expect and are looking for in the very near future. Your home will have a competitive advantage if some of the smarts are built in.

How Much Does a Smart Home Cost?

This is a question I'm asked often. Let's start with the costs of DIY smart homes. A DIY smart home is one built with smart home tech you buy from Amazon, Best Buy, or the like and install and configure yourself. Getting started with a smart home is pretty easy, but the costs can mount up. Since various levels of smart homes can do various things, we'll compare four scenarios:

- Bare-bones smart home cost
- Cost for expanding your smart home
- Full smart home build costs
- Adding luxury smart home items

For most DIY smart homes, a good home network with good Wi-Fi throughout your home is a requirement. I'm not going to directly consider that cost; however, depending on the size of your house, you will need a Wi-Fi router with a strong signal, and possibly an access point (AP) or two or a mesh system. This can easily run you $300–$400,[3] but in most cases, you aren't starting from scratch. Later in this book, I'll discuss more about making sure your home networking is up to par.

3 Here and throughout the book, all amounts are in United States dollars.

Bare-Bones Smart Home Cost

A bare-bones smart home is where most people start. It will have a handful of smart devices allowing you to use apps and voice to turn on and off a light and a switch, as well as a voice assistant for weather, timers, news, music, etc. A bare-bones smart home cost isn't too high:

- Smart bulb: $20–$70 (e.g., LIFX, Kasa, Geeni Prisma Smart Bulb)
- Smart switch: $20–$40 (e.g., Wemo, Kasa Smart Plug)
- Voice assistant: $40–$150 (e.g., Google Mini, Amazon Echo Dot, Google Home)
- Total cost for a bare-bones smart home: $80–$260

Costs for Typical Smart Home Expansions

After you've started your smart home, you're going to get the bug to expand on its capabilities. Most people enjoy the conveniences and capabilities of smart lights and smart switches, and start adding a few more. Smart locks are popular because people like the capability of remotely locking and unlocking their doors. They also like the comfort of knowing when their children come home and being able to check and see whether they locked their doors. Smart locks enable these functions.

The big step-up some gravitate toward is purchasing a hub to run and automate their smart home. Smart home hubs (e.g., Samsung SmartThings, Hubitat, or Vera) allow DIY smart home builders to integrate and automate their various smart devices. They also can provide secure remote access for controlling and monitoring smart home devices like your thermostats, lights, and security system. I use Home Assistant, a software hub, for my home. You'll learn more about smart home hubs later in this book. As you expand your smart home, you may incur the following costs:

- Smart hub: $0–$150 (Use a software hub for free, Samsung SmartThings, Echo Plus, etc.)
- Smart lock: $120–$250 (e.g., Schlage, Yale, August, and other smart deadbolts)
- Smart thermostat: $100–$250 (e.g., ecobee, Nest)
- A few more smart bulbs and switches: $150–$300
- Additional cost for extending your smart home: $370–$950
- Total cost so far: $450–$1,210

Full Smart Home Build Cost

Now you are fully into your smart home build. Most of the room lights are automated. You have a smart deadlock on your front door. But how about on your garage door entrance? Speaking of which, you'd like your garage door opener to be smart too.

Your kids love having voice commands, with Alexa telling them every answer. But having a voice assistant in one room is too limiting; let's add a few more in key spots so we have the voice command feature work everywhere!

It sure would be great to know who is at the door and when packages are delivered. Let's get a smart doorbell!

The kids make messes on the floor all the time and never want to clean up. I don't have maid service, but a robot vacuum will keep my floors clean!

There are so many things you can add, but costs also add up:

- Additional lights and switches: $100–$300
- Robot vacuum: $150–$1,100 (e.g., Ecovacs, iRobot, Neato)
- Garage door opener: $50–$250 (e.g., Chamberlain, NEXX)
- Additional smart lock: $120–$250
- Additional voice assistants: $100–$200

Complete smart home cost (Bare-bones ($80–$260) + Expansion ($370–$950) + Full Build ($520–$2,100)) = $970–$3,310

Icing on the Cake Costs

You're loving your complete smart home. But, despite all the money you've already spent and all the features you already have, you realize there is more you can build. Let's take it to the next level! You could add the following:

- Smart blinds throughout the house
- Connected weight/body composition scale
- Smart smoke detectors throughout your home
- Smart ceiling fans
- Leak sensors by the washing machine, water heater, and in the basement
- Smart refrigerator
- Integrated A/V systems
- Alarm system
- Smart lawnmower

- Connected irrigation system
- Smart display (e.g., Echo Show, Google Nest Hub)

You can also have a much bigger home requiring more hubs, smart switches, and smart lights, which can cost you more. In truth, the upper limit to what you can spend on a DIY smart home is very high. It all depends on what you want to do.

Your Labor and Time

A couple of thousand dollars is no small investment, but it may end up being a small investment compared to the time you spend on your smart home. Doing everything yourself, you'll spend hours researching the best devices, pricing, and compatibility. Then, after you've finally made your purchases, you'll have to install and configure each device. You'll do all the programming and automation, and all the fixing and troubleshooting. It can be a lot of work, but also a lot of fun, giving you a sense of accomplishment. Luckily, you have this book to save you some time!

Professional Smart Home Cost

What if you want a smart home but don't want to invest all that labor and time? You can have your smart home professionally set up by a Control 4, Savant, Crestron, or similar professional system dealer, but the costs are going to be much higher than the DIY costs. In addition to the equipment costs, you'll pay for installation and configuration, and you'll most likely need a maintenance contract, because you won't be able to make all the desired changes by yourself. Professionals will need to come to your house to make the changes (many of which you can, in fact, do yourself with DIY). That said, you save a ton of personal time and labor by not picking components, configuring, and troubleshooting.

So how much does a professional smart home cost? This is a very hard question to answer, because professional installer prices on equipment, installation, configuration, and options vary greatly. You can easily get up into tens of thousands of dollars and even into six figures.

This book is geared toward DIY smart home builders. I still recommend that you read the entire book, even if you are leaning towards a professional install. That way, you'll better understand the capabilities of smart homes. This will help you direct your professional installer to build the smart home you want.

Key Takeaways

- ☐ A smart home is a home that provides some combination of comfort, energy efficiency, security, lighting, etc., aided by technology that allows these systems to be automated, integrated, and made available for remote control.
- ☐ Reasons to want a smart home include the following:
    - ☐ Enhancing and enriching your life
    - ☐ Safety and security
    - ☐ Saving money
    - ☐ Fun and leisure
    - ☐ The cool factor

- ☐ Increasing your home's resale value

- ☐ Getting started with a smart home can be affordable.
- ☐ Smart homes can cost a lot of money and take up a considerable amount of your time.
- ☐ Professionally designed and built smart homes usually cost more than DIY smart homes and often don't give the owner full control over configuration changes.

A Day in the Life: Good Morning in Your Smart Home

Monday morning, at 5:45 a.m. – The smart lighting system in your bedroom slowly starts to brighten. Your blinds open just enough to let a little light in. This is all in preparation for your scheduled 6:00 a.m. wake-up time. The thermostat has just finished raising the temperature in your home to a comfy 72°F (about 22°C) after letting the temperature drop while you sleep to reduce energy usage.

6:00 a.m. – Soft music you've preselected starts to raise in volume slowly to get you up and out of the bed. You check your smart watch and notice you have a notification:

Your garden has been watered – Your irrigation system was scheduled to water the garden today. If there had been rain in the forecast, the irrigation system would have skipped watering today.

You take a shower and get dressed while listening to a curated news feed from your smart speaker.

6:30 a.m. – You head downstairs and the attractive aroma of freshly brewed coffee that your smart home prepared for you while you showered greets you. You pour a cup and enjoy your breakfast while continuing to listen to your news feed using your downstairs smart speaker.

7:00 a.m. – Your two children, woken up earlier by their own custom light and sound routines, come downstairs for breakfast. They each check the smart display on the refrigerator to see whether they have any activities. Your sixteen-year-old daughter, Katie, has basketball practice after school and isn't expected home until 5:00 p.m. Your eleven-year-old son, Micah, asks whether he can go to a neighborhood friend's house to play right after school. You tell him sure but that he has to be home by 4:30 p.m.

7:15 a.m. – It's officially sunrise, and the outdoor lights automatically turn off.

7:30 a.m. – The smart speaker lets everyone know it's time to head off to school and work.

Part 2

Planning Your Smart Home

Laying the Foundation

NOW THAT YOU KNOW WHAT a smart home is, why you want to build one, and how much it costs, it's time to start planning to have your dream smart home! In order to get your smart home off to the very best start, you need to plan for and consider the following:

- Make sure your home network, especially your Wi-Fi and security, is in good shape.
- Determine how much of a budget you have, including your time budget.
- Inventory the gadgets you already have.
- Make your short-, medium-, and long-term goals.

Building a Strong Home Network

It's best to make sure you have a fast, reliable, and secure home network before you get started building your smart home. Smart home problems can sometimes be tough to troubleshoot. Oftentimes, the problems people have with their smart homes are actually related to their home networks. Thus, it is best to make sure your home network is functioning as desired in advance.

Additionally, security is a major concern for home networks in general and smart homes that have Internet of Things (IoT) devices specifically. There are lots of very technical, jargon-filled definitions of IoT. Put simply, it is the concept of connecting any and every device in your home to the Internet. Traditionally, computers have been connected to the Internet in your home. Now, IoT devices like smart speakers, cameras, TVs, toys, wearables, and so much more are connected to the Internet. These devices typically come with fewer security features than traditional computers and expose your network to more vulnerabilities. This is why home network security is extremely important.

Router: Your Most Important Home Networking Device

Your router has an enormous impact on the security, speed, and capabilities of your network. Choosing your router is one of the most important home network decisions you'll make. In order to choose a good router for your needs, you should take the following steps.

Determine Your Bandwidth Needs

You primarily need to understand your Internet download needs. Download bandwidth needs are determined by the resources you pull into your network from the Internet (streaming services, gaming, torrenting, etc.). You want to make sure you have a router that can handle your Internet speed.

First, you should determine your current Internet bandwidth. Go to Speedtest.net[4] from a wired computer on your home network and test your Internet speed at different times of the day. Make sure no other devices are doing any significant downloading or uploading to the Internet (streaming videos, downloading large files, etc.) at the same time. Are they the speeds you were expecting based on the package you purchased from your Internet Service Provider (ISP)? If not, make sure to complain to your ISP.

Second, you should determine your bandwidth needs. For example, Netflix recommends 5 Mbps for each high-quality HD Stream and 25 Mbps for each Ultra HD stream. Below are some common bandwidth needs for home network activities:

- HD Video streaming: 5–25 Mbps per stream (downloading)
- Music streaming: < 1 Mbps per song (downloading)
- Online Gaming: 3–6 Mbps download / 1–3 Mbps upload per person
- Web/Social Media: 3 Mbps download / 1.5 Mbps upload per person
- HD Video Call (e.g., Zoom): 1–4 Mbps upload/download per person

The amount of bandwidth your home needs depends on the number of people and their regular usage of the Internet at home. For example, let's say it's common in your home for

4 https://www.speedtest.net/.

a couple of people to be streaming a video (~15 Mbps), for two more to be browsing the web/social media (~6 Mbps), and for someone else to be gaming online (~6 Mbps) and streaming music in the background (~1 Mbps)—all at the same time. Adding up 15, 6, 6, and 1 gives you up to 28 Mbps, which is the minimum download bandwidth needed to support these activities. You should always add at least a 30 percent overhead to your calculated minimum. The speeds ISPs quote are maximum figures that aren't always available. Often, there are other devices on the network that are steadily using small amounts of Internet bandwidth (e.g., checking for and downloading updates). Keep in mind that if you frequently view and/or send videos, pictures, and other files via social media, your needs may be higher than the average 3 Mbps figure above. Also, there are often spikes in the rates needed by these services. For example, an online game may average 3 Mbps but have certain spikes where it needs 10 Mbps to function properly. In the example I gave earlier, I'd recommend you to actually get around 40–50 Mbps to cover your needs.

If you constantly download content (via torrents and/or Usenet, for example), then you already know that you have high bandwidth needs. Ultimately, it comes down to how fast you want those downloads to happen and how often you download.

You may wonder about upload bandwidth. Examples of activities that require more upload speed include VPN connections, video calls, and some online games. However, upload speeds from Internet providers are usually the same or slower than their download speeds. Your router will most likely support your upload speed if it supports your download speed.

What I Do

For a while, I had a Comcast/Xfinity package that was 140 Mbps down and 10 Mbps up. That was okay for many years. However, as my family cut the cord and got rid of cable, we started streaming video a lot more. Although 140 Mbps bandwidth was fine for our streaming needs, Comcast had a 1 TB monthly limit and would charge you more if you passed that limit. I eventually switched to CenturyLink fiber service, which has speeds of 1 GB up and down without a data cap. And to top it off, it's less expensive than what I was paying Comcast.

Determine How Much Space You Need to Cover

Most people use their router as a wireless access point, which is a device that allows Wi-Fi connections to your home network. If you do, you'll want a wireless router that provides your entire home with stable, fast Wi-Fi. Consider at least two factors when determining the range:

1. What protocol(s) do you plan to use? Some, like 2.4 GHz protocols (e.g., 802.11n, 802.11g), cover more range and support more devices. Others, like 5 GHz protocols (e.g., 802.11ac, 802.11n can be 5 GHz as well), cover less range but are faster and less prone to interference. The latest protocol is Wi-Fi 6 (802.11ax), but it will be a while before most devices adopt it.

2. Where in your house will you place your wireless router? Generally, the best place for whole house coverage is in the center of your house, but few people place a router

there. If it is at the edge of a floor or house, it has less chance of covering the house.

Wi-Fi range for routers varies greatly. A good Wi-Fi router will cover 2,000 square feet easily. Some can cover up to 5,000 square feet. Remember these three key factors that can impact Wi-Fi range:

- **Signal Interference** – If you live in an area flooded with Wi-Fi signals (e.g., an apartment complex), you'll have interference that restricts your Wi-Fi signal stability and range. Microwaves, cordless phones, Bluetooth speakers, and even baby monitors can also cause signal interference.
- **Walls and surfaces** – Materials like concrete, metal, and plaster are the worst for blocking Wi-Fi signals.
- **Router placement** – I mentioned earlier that router placement affects signal range. Most routers are omni-directional. Placing them next to concrete or brick walls will limit their range. Also, the higher up on the wall the better the range of the signal.

If you have a large home, one or more factors limiting your Wi-Fi signal range, or lots of devices, you should consider buying a router and supplementing it with additional wireless access points. Companies like TP-Link and Ubiquiti sell wireless access points specifically made to expand wireless coverage. Another option, especially if you don't have wired network connection for your wireless access points, is investing in a mesh router system (e.g, Google Nest Wi-Fi and Eero).

Determine Budget

I recommend that you budget at least $100 for a good Wi-Fi router. You can get away with less. However, if you can get to the $100 mark, then there is a significantly lower chance you'll be disappointed with your purchase. I'd recommend saving up until you can reach that point, because the wireless router is an important piece of equipment for your network. If you are willing to go up to $200, you should be able to get all the features you are looking for.

Pick a Wireless Router

Seems simple, right? Here are a few things to look for and some to avoid:

Features to look for:

- ☐ Support for Wireless AC. There's no point in buying a router that doesn't support this protocol. It's faster and more stable than the protocols that preceded it, and most mobile devices made in the last few years support it. Wi-Fi 6 would be a bonus and would help you in the future.
- ☐ Supports MIMO. MIMO (multiple-input and multiple-output) allows the radio in your router to connect to multiple devices without slowing down each connection when compared with wireless routers without MIMO. SU-MIMO (single user) is good, and MU-MIMO (multi-user) is even better.

- ☐ Quality of Service (QoS). This allows your router to specify the types of Internet traffic (e.g., video streaming, gaming, etc.) that have priority over other types of traffic. QoS can also be used to give priority to certain devices (e.g., your mobile phone or PC) over other devices.
- ☐ Supports WPA2 AES security for Wi-Fi. Any older standard isn't secure.
- ☐ Gigabit Ethernet ports. The wired connections to your router should also be fast.
- ☐ Guest Network. A guest network is an essential part of Wi-Fi security. I'll explain this in more detail later.
- ☐ Consistent firmware updates. You should be able to set up automatic security updates or at least be able to be notified or check for them from the router interface.

A Couple of Things to Avoid

- ☐ ISP's provided model. Chances are this model is bundled with a modem, making it more of a single point of failure from a security and reliability standpoint. Renting this unit over time will probably cost more than buying one, and these routers are often fraught with security issues.
- ☐ Older routers. Get one that is new or at least receives frequent and recent security updates.

Read Reviews Before Purchase

You may have narrowed it down to two or three routers. Use professional reviews (CNET, Tom's Guide, or my site,

HomeTechHacker[5]) and user reviews (Amazon, Newegg) to get opinions and learn more about the router to help you make your decision. This is the way to get real-world answers if you have questions about the router.

What I Do

A few years back, I became somewhat frustrated with off-the-shelf routers. There seemed to be many security vulnerabilities and not enough firmware updates. I decided to build my own router using pfSense software. pfSense is free for home use and has tons of features that are found in enterprise-class routers. It has all the features I need, is very secure, and has served me well for many years.

These days, there are better choices for routers, and pfSense isn't for everyone. However, if you are feeling a little adventurous, give it a shot.

Switches and Cabling

With where technology is going, you should try to achieve at least gigabit speeds in your home. The more bandwidth you have in your home, the more future proof it is and the less chance there is for bottlenecks. You can achieve this by making sure all wired networking equipment (switches, routers, access points) and all wired clients (desktop computers, laptops, streaming devices) support gigabit networking. It's not the end of the world if some devices support only fast Ethernet (100 Mbps). That's plenty fast for most uses. But you want to be prepared for the future.

---

5 https://HomeTechHacker.com/.

You can support gigabit speeds and make your home more compatible with future speed technologies by using at least Cat 5e cabling and preferably Cat 6. Cat 5e/6–rated cable will support gigabit speeds for longer distances and more reliably than older standards like Cat 5. While recabling a home can be hard, if you have the opportunity to choose your cable, go with at least Cat 6 or even Cat 7.

One other consideration for the wired part of your home network is that you should use as many wired devices as possible. Although Wi-Fi standards keep improving, the fact is that wired is generally more reliable. Try to save your wireless bandwidth for devices that truly need it (e.g., mobile devices and devices that don't have wired networking).

Make Sure Your Wi-Fi Network Is Rock Solid

Even though I prefer wired over wireless connections, I realize that most smart home devices are wireless. Have you noticed how important the Wi-Fi connection in your house has become? With the rise of mobile devices and the IoT, your home Wi-Fi is more important than ever. Cell phones, streaming devices, smart bulbs, home security systems, washers and dryers, smart speakers, TVs, and so many other gadgets now require Wi-Fi connections in order to access all the geeked-out features that caused you to buy them in the first place.

You've probably already developed a major reliance on your Wi-Fi. Making your home smart will only increase that reliance, so you need to make your Wi-Fi fast and stable. Here is how to achieve both goals.

Making Sure your Wi-Fi Network Is Fast

Many factors contribute to the speed (or lack thereof) of your Wi-Fi network. Let's walk through the major factors and how you should address them.

1. **Faster Wi-Fi protocols** – At the time of writing this book, Wi-Fi 6 (802.11ax) is just starting to become widely available. For faster and more reliable speeds, your Wi-Fi network should support the faster 5 GHz protocols like Wi-Fi 5 (802.11ac) and 5 GHz Wireless N (802.11n). A lot of devices still use only 2.4 GHz protocols like 802.11g, which your network will still need to support. However, I recommend that you move as many devices as you can over to the 5 GHz protocols, which are faster and less subject to interference.

2. **Number and types of devices connected to your Wi-Fi** – How many wireless devices do you have on your network? Do you stream pictures and videos wirelessly across your network? Do you run some type of networked home surveillance system, or do you have some wireless cameras in your home? These are the major culprits hogging your wireless bandwidth. You want to make as many of these devices as possible wired clients. Also, just having too many devices for your router to handle can cause problems. Most routers will tell you they can handle way more clients than they actually should. How many clients they can handle without problems really depends on what those clients do, how often they do it, and whether they do it concurrently. You may need to add some access points to distribute the load

and also to cover Wi-Fi dead spots in your house. (More on this in the next section). I shoot for no more than forty wireless devices per wireless router/access point.

3. **Internet Speed** – I mentioned this earlier in the book, but you want to make sure you have enough overall Internet bandwidth for your home network. With the exception of video streaming devices and network cameras, most smart devices don't need much Internet bandwidth. Your Wi-Fi speeds will always be constrained by your Internet speed when accessing cloud and other services on the Internet.

Make Sure Your Wi-Fi Network Is Stable

Having stable Wi-Fi in your home means having a strong signal everywhere you need it. It also means your signal doesn't cut out from time to time and can handle the demands of all of the wireless devices on your network.

Most people rely on their router's wireless access point capabilities for their wireless network. This can be cost effective, especially for smaller homes that need only one wireless access point to provide coverage for the entire house. However, lack of flexibility with where you can place the router can be a problem. For most homes, the ideal place to put a single wireless AP is somewhere near the middle of your home high up on a wall or on the ceiling. This spot is usually not a reasonable option because routers need to be near the modem provided by the ISP, which is often in a room nowhere near the middle of the house. Also, many people have a modem provided by the ISP, which also functions as the router and AP, further constraining its placement.

If you can't place the router in a good spot in your home or have a larger home that a single router can't cover, then you will have spots in your home that aren't covered or have poor wireless performance. You can fix this by buying a Wi-Fi range extender, a wireless access point, or by replacing your router with a mesh system. Let's take a quick look at these options.

A range extender simply rebroadcasts your wireless signal to other parts of your home, thereby extending your wireless coverage range. Some have extra features like providing an Ethernet connection and providing a guest network. The downside of range extenders is that there will be some speed loss, but some Wi-Fi extenders greatly reduce this loss by using multiple radios (separating the duties of receiving and transmitting the signal) and faster processors. Look for a range extender with these features.

A wireless access point, despite its name, plugs into a wired port somewhere in your home and can broadcast a wireless signal from there. Unlike range extenders, there is no wireless speed loss because they aren't rebroadcasting a signal. The downside, when compared to Wi-Fi range extenders, is that they need access to an Ethernet connection to work. Many also have features like guest networks and traffic monitoring. If you have access to a wired connection in an ideal spot in your home, this is my recommended option. If you have a really large home, you can use multiple access points.

Mesh Wi-Fi systems (e.g., like the Netgear Orbi or Google Nest Wifi) have become really popular. You can think of them as multiple easy-to-setup range extenders that are made to work together. Most systems give advice and tools on how to place them to get full coverage for your home. One key piece of information to know is that mesh systems are also designed to replace your router. This means they come with tons of features that Wi-Fi extenders will not have. If you don't want to or can't replace your router, then mesh systems aren't for you (unless you want to get into the complications of running multiple routers on your network, or you get a mesh system that can work in bridge mode). This is the most expensive of the three options. If you have a large home that doesn't have wired networking throughout the house, this is probably your best option.

Be sure to watch out and plan for being able to support the sheer number of wireless devices on your network. Theoretically, most wireless access points can have 255 devices connected at a time, but that doesn't mean they'll work well with that many. You probably won't find the maximum limit in your access point's documentation. If you have only one access point, I'd start to worry about having too many Wi-Fi devices when you get to around 20–25 devices connected at a time. At this point, depending on how active those devices are, you may also start to run into bandwidth issues. I mentioned how to address bandwidth issues earlier, but if you believe you have fixed your bandwidth issues and are still experiencing slowness, you may want to upgrade your access point or add an access point to distribute the load of your wireless devices.

Some of you may be thinking that you'll never get to 20–25 wireless devices. It can happen faster than you think. Smartphones, tablets, streaming sticks, video game consoles, smart TVs, smart hubs, smart bulbs, thermostats, smart plugs, smart refrigerators, smart toothbrushes, smart scales, smart watches, smart smoke detectors, smart washers, dryers... I think you get my point.

What I Do

I have prioritized making my wireless network fast and stable. My router doesn't have wireless functionality, so I rely completely on wireless access points. To make sure I have good coverage throughout my home, I have placed a wireless access point on each floor of my home. I'm currently using relatively inexpensive TP-Link EAP225/245 access points, which support both 2.4 GHz and 5 GHz (802.11n/ac) signals and provide plenty of speed. I have about seventy devices on my Wi-Fi network, including surveillance cameras and streaming devices, but I don't notice any speed or stability issues.

One really nice thing about the TP-Link access points is that I can manage them centrally using free controller software (Omada) provided by TP-Link. This allows me to create and manage the wireless network and connected clients from one spot or using an app. I can see which devices are connected to which access point and the strength and speed of their connections.

Secure Your Home Network

Cybersecurity should not be an afterthought for any home network implementation. You probably have more valuable information on your home network than you realize. Private family pictures and photographs. Sensitive financial information. Important software and devices that you and your family count on every day. If someone breaches your network security, they can compromise all of these things. If you are going to build or expand your smart home, you are increasing the number of potentially vulnerable devices that you rely on. Have you taken the steps below to protect your home network? Here are some basic, intermediate, and advanced tips you should consider for securing your home network.

Basic Security Steps

These steps you can easily do in ten minutes or less. They don't require much technical knowledge, and I recommend that you implement all of these immediately.

Update the firmware on your router

Your router (and the firewall in it) is your primary defense against hackers penetrating your network. Having an old router and/or a router with old firmware on it is a big security risk. New vulnerabilities are discovered all the time, and you want to make sure that your router has the latest protections. Routers are set-and-forget-type devices for most people, but if your router is no longer receiving updates, you should consider replacing it, as it may be too old.

Make these changes to your router settings

This may seem basic, but make sure you log in to your router and check the following:

- Remote access (from outside your network) is disabled. You want to restrict the ability to log in and change your router settings. You want to be able to configure your router from your home network only (or even from only specific computers, but that's a more advanced topic).

- Your firewall is enabled. This should be the default, but you'll want to make sure!
- You have a strong password for logging into your router. Change the username from the default if you can.

Change your home Wi-Fi password

Please tell me that your Wi-Fi password isn't the default one (and that you do have a password and security turned on for your Wi-Fi network). Implement a strong password for your Wi-Fi network, and change it every so often, especially if you have neighbors that are in range of your Wi-Fi network.

Keep your Antivirus/Firewall programs up to date

While your router firewall is your primary defense for your network, a strong antivirus and firewall program running on your computer is the primary defense for your computer. Most malware comes from sites users interact with, not from hackers getting into your network. Protect your computer by making sure it has the latest protections your firewall and antivirus program provide.

Intermediate Security Steps

This is the next level of improving the security of your home network. These may take a day or so to research and set up, but they don't require deep technical knowledge.

Encrypt your hard drive

This is especially important for mobile devices like laptops, tablets, and mobile phones. If these are lost or stolen, thieves can easily access the hard drive without knowing the login information for your device. Encrypting the hard drive makes it much more difficult for thieves to access your sensitive data, even if they have physical access to your device. If encrypting your entire hard drive seems too drastic, or like too much work, you can encrypt just the sensitive files you want to be protected.

Set up regular backups

Although not a direct protection against a network breach or computer infection, having a backup strategy can save you from a catastrophic loss of data if there is a breach, or even if hard drives or computers fail. There are many options for backup, including backup to a separate hard drive and remote backups.

Set up a guest network

You need to protect your network from guest devices (in case, they are compromised) and not give out your regular Wi-Fi password to guests. Make sure you set up a network just for guests, which is isolated from your regular network.

Advanced Security Steps

These steps, for many, will take time and effort to design and implement. In my opinion, they are worth the time. You may need technical help in implementing these strategies.

Secure your network from IoT devices

If you have not segregated your IoT devices from the rest of your network, you are at the mercy of notoriously insecure devices (webcams, smart TVs, smart bulbs, etc.) remaining secure. There are a few different ways to do this, and some of these ways are very difficult and take some maintenance. A simple way is to put all your IoT devices on your guest network, but this can cause complications if other devices on your network need to interact with your smart devices. I have some more tips for isolating your IoT devices. See the end of this section for details on what I do for home network security.

Monitor your network for unknown/new devices

Do you know all the devices that are connected to your network? Do you occasionally check your network to see whether unknown devices have connected? Ideally, your router allows you to monitor the devices connected to your network. It would be even better if your router could alert you when new devices connect to your network. Most routers don't have that function, but you can use software that scans your network for devices and sends alerts. There are free and paid options for whatever OS you run.

Monitor your logs for breaches

This takes network monitoring one step further. Network professionals often aggregate the logs of all their machines in one system for analysis. Part of their analysis is to detect intrusions. There are many free options and paid options out there for doing this, such as Splunk, Logagent, and even Rsyslog. These tools can send you alerts when there are too many password failures, unknown machines on your network, and many other security concerns.

Implement malware prevention and detection at the network level

You can actually process incoming Internet traffic for your home network and filter out malware and known hackers. This is another feature of some higher-end routers and router software (e.g., pfSense). There are various software packages for intrusion detection systems (IDS) and blocking known "bad actors" from accessing your network. The easiest way to do this is to buy a router that has these features.

What I Do

As I mentioned earlier, my first line of defense is my pfSense firewall and router. pfSense is actively maintained, and I make sure I keep up to date with new firmware. I've made certain that remote access to my router is disabled. I've also taken additional steps to ensure only a couple of computers on my LAN can access my router's user interface, for additional security.

Snort and pfBlocker are packages available with pfSense, which I use to further protect my network from bad actors. Snort detects known intrusion signatures—telltale signs that show a system is being attacked—and blocks hackers trying to access my network. pfBlocker takes a different approach by reading regularly updated lists of known computers of hackers and other bad actors and blocking them from accessing my network.

I have more computers in my house than most because I repurpose old hardware into file servers, multimedia distribution systems, virtualization hosts, etc. Every single one of these machines runs a firewall with rules to block other devices in my house from accessing popular network services (e.g., web connections, file and printer sharing, remote desktop, etc.). These rules are customized for each machine. My file server obviously allows most computers to access its files, and my web servers allow HTTP/HTTPS connections. By default, however, my firewall rules are set up to block all my IoT devices from accessing anything unless I write in an exception. All of my computers run Linux, and I use the UFW firewall software to accomplish this.

For additional Wi-Fi security, my home is separated into three Wi-Fi networks: home, guest, and IoT. Each Wi-Fi network has a uniquely identifying name. Computers on my home network have access to all other computers on my network, and security is a bit laxer. Computers on my guest network cannot access any other computer on my network and are also restricted from hogging all of the bandwidth on my Wi-Fi. Computers on my IoT network are restricted from accessing most computers on my home network, with some exceptions. For instance, my security cameras need to save their video to my surveillance system hard drive, so they have access to that machine. My home automation controller (Home Assistant) needs to control many of my IoT devices, so they are permitted to access each other.

I put in an additional restriction for my security cameras. I use firewall rules in pfSense to block them from accessing the Internet. This way, hackers cannot look at the video from my cameras, and my cameras can't send video to the Internet, only to my home network.

For backups, I employ the popular "3-2-1" backup strategy which stands for the following:

- 3 copies of your data
- 2 local copies on 2 different devices (original and backup)
- 1 offsite backup

I have a home-built fileserver, and I back up all important documents, photos, videos, and other important files to a separate machine from my fileserver nightly. SpiderOak is the cloud backup service I use to remotely back up my fileserver. All the computers in my house have access to the file server. My family knows that their important files are backed up locally and remotely every night.

Graylog2 is the software package I use to centralize all of the logs on my network. I have all the system logs from each computer, my router, and my access points sent to my Graylog service for processing and storage. Graylog is set up to alert me when unknown computers join my network (wired or wireless), when there are too many failed login attempts to any machine, and even when there are power outages (it receives information from my uninterruptible power supply). Additionally, from one interface I can search through the logs of all of my machines when I am trying to hunt a problem or an intrusion down.

Know Your Budget and Time Availability

Building a smart home can take a little time or a lot of time, depending on what you want to accomplish. It can also cost a little money or a lot of money. You should know how much time and money you plan to spend on your smart home. These commitments will help determine what you can reasonably expect from your smart home–building efforts.

I wrote earlier about the costs of a smart home. To summarize,

- bare-bones smart home costs from $80 to $260;
- cost of expanding your smart home adds between $370 and $950;
- full smart home build adds costs from $520 to $2,100; and
- adding luxury smart home items depends on how much you want to spend.

All in, a full buildout of a smart home from scratch will probably cost you from around $1,000 to $3,500. You can get by with spending less, and you can certainly spend more. Most people will be very happy with anything built in the $1,000–$3,500 price range. Remember, you don't have to (and probably shouldn't) buy it all at once. That's the cost over time as you accumulate devices. You should decide how much you have to spend before you get started.

It's important to consider the costs before diving in and to remember the cost of your personal time. Luckily, you don't have to break the bank to get started, and you can gradually add items over time (which is a good thing, since it takes time to configure each device). A smart home will be a labor of love. Deciding whether it's worth the cost of equipment is easy. Deciding whether it's worth your time and effort takes more thought.

Make Short-, Medium-, and Long-Term Goals

Building a smart home should be treated like other home improvement projects. Plan your goals and timelines before you get started. These goals will help guide your purchase decisions and priorities. Although this section is titled "short-, medium – and long-term goals," feel free to group your goals differently. The point is to have goals at different stages, because you are probably going to build up your smart home in stages.

Short-Term Goals

I would define short-term goals as the goals you have in the first six months to a year of building your smart home. If you are completely new to building a smart home, one of your primary goals for the first year should probably be to learn about smart homes and smart home technologies. Another goal should be to spend some time thinking about tasks you'd like to automate or cool effects and gadgets that you want to integrate into your home. Another great short-term goal is to familiarize yourself with what other people are doing with smart homes, for inspiration. Check out the smart home buildout examples in the Appendix for inspiration and ideas.

Your short-term goals should also include what you are going to implement in the first year. Don't be overly ambitious with your implementation goals when you first start out. If you go too fast too soon, you may end up really frustrated. Worse yet, you could end up redoing a lot of what you first implement as you learn more. Pick a few simple things, enjoy them, and learn from them to help you set and revise your medium – and long-term goals.

Medium-Term Goals

Medium-term goals are what you want to accomplish over the first one to two years. Hopefully, you've learned enough over your first year of implementation to be able to more specifically envision how you want to accomplish your medium-term goals. These goals can be a lot more ambitious as long as your time and budget are sufficient. You should feel more confident in tackling more advanced projects.

Long-Term Goals

Long-term goals are less about a time period and more about the state you want your smart home to be in. Your long-term goals should accomplish all that you've dreamed that your smart home can do. It may take you a few years and quite a few dollars to get there if you have ambitious goals. Others will be able to accomplish all of their long-term goals in the first year!

Remember, these goal terms are just guidelines. Everyone starts from a different level of comfort and knowledge, and what they've already installed when planning their smart home. The most important point to take away is that you should set reasonable goals, chunked out over reasonable durations, and revise those goals as you learn more.

What I Do (Did)

When I started out building my smart home, I set the following short-, medium-, and long-term goals:

- Short-term:
    - Reduce energy usage by automatically turning off lights that aren't in use.
- Medium-term:
    - Learn more about smart home technologies and pick the right smart home hub to control a few more lights and door locks.
    - Learn to write some automations.
    - Monitor and improve home energy usage.
    - Pick a home security system that integrates with my smart home.
    - Set up video surveillance.
- Long-term:
    - Set up remote control of all of my lights.
    - Automate my Christmas lights.
    - Integrate with Alexa and Google Assistant voice control.
    - Control and monitor my A/V equipment.
    - Set up smart control of my HVAC (Heating, Ventilation, and Air Conditioning) systems.

- Integrate all of my smart home devices.
- Keep thinking of things to automate.

I've learned a ton over my smart home journey, and I keep adding goals and planning future projects. Hopefully, this will happen to you as well.

Key Takeaways

- ☐ A strong home network is the foundation of your smart home.
- ☐ Make sure your Wi-Fi network is fast, stable, and secure.
- ☐ Secure your home network from bad actors.
- ☐ Make short-, medium-, and long-term goals for your smart home build.
- ☐ Set a dollar and time budget.

A Day in the Life: Your Smart Home while You're at Work

8:00 a.m. – You're just arriving at work when you get a notification on your smart watch:

No one is home and the alarm isn't armed – In the morning rush to get to work and school on time, everyone forgot to arm the alarm. You press a button on your smartphone that arms the alarm. Since no one is home, the smart house also turns off any non-security lights left on, makes sure the doors are locked, and ensures the garage doors are closed.

8:15 a.m. – Your smart thermostat senses that the house is empty and stops actively maintaining the comfortable 72°F (about 22°C) temperature. It will regulate the temperature if it gets too hot or too cold, but otherwise it conserves energy.

10:00 a.m. – The robot vacuum cleans the kitchen and dining room floors. It sends you a notification that it's time to replace one of the robot's brushes that sweeps the floor.

12:00 p.m. – The robot lawnmower mows the lawn.

4:00 p.m. – The thermostat starts returning the house to the desired comfortable temperature in anticipation of family members arriving home.

4:36 p.m. – You get an alert on your smart watch that Micah has arrived home. The smart home knows that Micah is home, because he has his own combination to the smart lock and the alarm system. He's a little late, but home safely.

5:00 p.m. – You get an alert on your smart watch that Katie is home from basketball practice. She has her own codes to the alarm and smart locks too but didn't use them because Micah was already home. But she has a smartphone that alerts the smart home to her presence.

Part 3

Building Your Smart Home

Key Components of the Modern Smart Home

THE TERM *SMART DEVICES* MEANS many different things to many different people. In this section, I'm going to enumerate and describe the uses and capabilities of common smart devices that are in the modern home. This will help you to plan the various stages of your smart home build by knowing what different types of devices can accomplish and inspire your goals.

Smart Lights

Smart lighting is where a lot of people start when building their smart homes. They have been available and easy to use for years. The immediate visual payoff is alluring to consumers. There are all kinds of smart lights out there, including these:

- **LED Bulbs** – These are bulbs like the ones that are already in most houses. However, they add features like changing colors and brightness. You can schedule them to come on and go off when you want and even use an app to control them without getting up from your chair.
- **LED Strips and Shapes** – These are products that allow you to add lighting in places where you don't have a light fixture. You can add backlighting to your TV, cabinets, bed, bar, and many other places.

Smart Switches and Plugs

Another popular first product for new smart homes is smart plugs. These can plug into or replace existing power outlets and turn them smart. They allow smart home users to

- turn devices on and off remotely (via an app or voice assistant),
- schedule when devices turn on and turn off,
- monitor how much power a particular device consumes.

Smart switches are very similar to smart plugs. They replace (and in some cases work in conjunction with) existing light switches and allow you to remotely control your lights. Smart switches have some advantages over smart bulbs:

- Smart bulbs don't work if someone turns their power off at the light switch. For any scheduling or other automations to work with smart bulbs, the light switch must always be on. Smart switches, on the other hand, can be turned on and off remotely.

- Smart bulbs can add up in costs. It's much less expensive to just use one smart switch instead of replacing all the bulbs if you want to automate lighting in a room with multiple light bulbs connected to a single switch.

The disadvantage to smart switches is that you'll have to get comfortable with standard home switch wiring to install one, or you'll have to add on the cost and hassle of having an electrician do it for you. Additionally, smart switches don't make regular light bulbs change colors. You can, however, get smart dimmers that can dim compatible bulbs.

There are other types of smart switches that control more than just light switches. For instance, there are smart fan controls that are wired into the wall or the fixture that provide a smart switch for the fan's speed and light.

Smart Locks and Garage Door Openers

In addition to being convenient like smart bulbs and smart switches, smart locks add security to your home. The best security feature of smart locks is that you can remotely control, monitor, and automate the locking of your doors. Ever wondered whether you locked your front door after you've left, or closed your garage? Smart locks and smart garage door openers can tell you and even allow you to lock/close them in case you forgot. Smart locks also have the following benefits:

- You no longer need a key. You can simply remember a code or even use a fingerprint or your smartphone to open your lock.
- You can set up alerts for when the door is locked and unlocked.
- Each individual in your home can have their own unique identifier (fingerprint, phone, code) for entering the home. For parents, this means you can know when your children come and go.
- You can set up temporary guest codes for people coming to your home.
- You can revoke access to someone or change anyone's code at any time.
- You can schedule your doors to lock whenever you want. For instance, you can schedule your doors to lock each night at a specific time in case you forget to lock them manually or fall asleep watching TV. This way, you know your home will always lock up for the night.

Smart garage door openers have many of the same benefits as smart door locks. I've found that it is just really nice to know the status of my garage door and be able to open and close it remotely.

Smart Sensors

Smart sensors give you a lot of information about what is going on in your home. This information ends up becoming the backbone of automating your home and making it truly smart. There are so many different types of smart sensors.

Here are some of the more popular ones and some of the ways to use them:

- **Motion Sensors** – These are great for security systems and for knowing when someone is in a room or home so you can turn the light on or adjust the home temperature.
- **Door/Window Sensors** – These sensors can tell you when a door or window is open or closed. This is great for security and for knowing when your kids are getting into the candy cabinet.
- **Temperature Sensors** – Thermostats are great, but they are not always well positioned in the house. Temperature sensors can tell you the temperature (and often humidity) of wherever you put them. This can help you to automate your HVAC system.
- **Air quality sensors** – These can give you an idea of the air quality in your house and automate turning on air circulation and/or alerting you to danger.
- **Smoke/CO/Natural Gas Sensors** – These also can alert you (even remotely) to danger in your home.
- **Light Sensors** – Light sensors can determine how much light is in a room or area. You can use these to more accurately automate turning lights on or off or changing the brightness of lights.
- **Water Leak/Flood Sensors** – These sensors can alert you when water is where it isn't supposed to be (like a flooded basement).
- **Vibration sensors** – Having a sensor that can tell when something is vibrating is useful for knowing when a washer or dryer is done, or when someone

breaks a window. I use one for determining when my generator is running.

- **Multipurpose Sensors** – Multipurpose sensors do one or more of the above. For example, I have a few multipurpose sensors that detect motion, temperature, humidity, and light levels. I also have a flood sensor that also detects the temperature.

Smart Thermostats

Smart thermostats are another common way people enter into smart homes. They are similar to smart sensors in that they detect the temperature and sometimes air quality and humidity. They take it one step further and control your HVAC system. They allow you to remotely adjust your temperature. They can learn your patterns and adjust the temperature for you.

Many of them give you access to stats and suggestions about the efficiency of your HVAC system. This, along with smartly adjusting your temperature based on home occupancy and comfort, can lead to cost savings on electricity and gas bills.

Some people find smart thermostats difficult to install. Usually, the difficulty is dependent on your HVAC system and the wiring in your home. A licensed electrician will have no problem installing one for you.

Smart Security and Surveillance

Home security systems and home surveillance cameras have been around for decades. Smart home security systems take things to the next level by integrating with other smart home devices like door locks and lights to further secure your home. This integration allows security systems to flash your lights when an alarm has been set off and to lock doors when the alarm is armed.

Smart security cameras take a few different forms. There are IP (Internet Protocol) cameras that stream video over the network. There are also smart doorbells that can notify you when someone presses a doorbell and allow two-way video and/or audio conversation with the person at your door, even if you aren't home! Both types of cameras can have advanced features like facial recognition, which can let you know who is at your door or looking at your camera.

It's not too hard to put your entire home into a surveillance state. Whether or not you should do that is well beyond the scope of this book.

Smart Remotes and A/V Equipment

These days, regular multimedia and audio/video (A/V) equipment can reasonably be considered smart. Most new TVs come loaded with smart features (e.g., access to popular streaming services like Netflix and Amazon Prime Video). Some also allow for remote control via an app or voice assistant. A/V receivers offer many of the same features, sometimes replacing video streaming services with music streaming services like Spotify. Then you have all of the streaming devices out there. Roku and FireTV devices dominate this market. Both have many smart integration features with voice assistants and apps.

In addition to smart streamers and A/V equipment, there are also smart remotes like the Harmony series. Some of these remotes come loaded with features and capabilities that allow them to control smart devices like smart switches, smart lights, and smart plugs as well as the A/V equipment they have traditionally controlled.

Smart Home Hubs

A smart home hub integrates and facilitates the automation of the various smart devices in your home. Most smart devices can't communicate directly with each other. Smart home hubs broker communication between smart devices, which increases the flexibility and control of your smart home.

There are many types of smart home hubs. They are essential to a smart home. We'll go over in detail how to pick the right one for your smart home later in this book.

Smart Speakers/Voice Assistants

Smart speakers and voice assistants are everywhere. They are in speakers, our phones, appliances—the list goes on and on. Alexa and Google Assistant dominate this market, but there is also Siri for Apple fans. These devices do much more than answer questions and tell jokes. Voice assistants can function as smart home hubs because they can control smart home devices. They have become a ubiquitous part of smart homes.

Other Smart Devices

I can't possibly list all the different types of smart devices, because more and more are envisioned and built every day. Robot vacuums, robot lawnmowers, smart scales, smart irrigation controllers, smartwatches, and many more devices are showing up on shelves at a rapid rate. Sometimes, it seems as if every device will eventually be turned into a smart device.

Getting Started Building Your Smart Home

Now that you have your infrastructure, plan, goals, and knowledge of smart devices in place, it's time to start building. Setting up smart homes used to be extremely difficult and only for the technically inclined. Today, I believe the technology has advanced enough to let anyone build a smart home. Provided you've absorbed most of the earlier material in this book and follow the steps below, your smart home will be up and running in no time.

Inventory Your Home for Smart Devices

This is a step most expert advice fails to address and most people forget to do. There is a good chance you have a few smart home devices already if you are reading this book. Ask yourself the following questions (you can also use the checklist in the Appendix):

- Do you have a voice assistant (e.g., Alexa or Google Assistant)?
- Do you have any smart lighting? Maybe a timer switch for lights? Even old-school mechanical timers provide smart home capabilities. They are

just difficult-to-impossible to integrate with other smart home devices.
- Do you have a TV or A/V receiver that can be accessed over the network?
- Do you have a robot vacuum?
- Do you have an alarm system? This system may have smart capabilities.

Inventory the devices you already have to see how they may fit in your smart home goals. You are probably not starting from scratch. When I started making my home smart in earnest, I already had timers for my outdoor porch and Christmas lights. I also had some motion sensor light switches. These worked well, and I didn't replace them because I didn't need to. Later on, you may want to replace some of them so they integrate better with your other smart home devices. But for now, it's good to know what you have.

Test the Waters

Now, it's time to implement the first parts of your new smart home. It's actually not that hard or expensive to get started. Let's go step by step:

Step 1: Buy a Smart Bulb

Smart bulbs are a good first foray into the smart home because they can be extremely useful and fun, are easy to integrate, and can be relatively inexpensive. You can program them to turn on and off at certain times of the day, and eventually integrate them with motion sensors to turn on automatically. You can also create moods and custom atmospheres with lighting and color effects. I recommend getting a multicolor smart bulb (e.g., LIFX (A19) Wi-Fi Smart Bulb). The Philips Hue smart bulbs are just as great but require you to purchase their hub, making them more expensive.

Step 2: Buy a Smart Plug

In addition to lighting, another key feature of a smart home is being able to automate and remotely turn on and off devices, using a smart plug. You can remotely turn on or schedule your coffee maker. Also, you can automate Halloween or Christmas lights, lamps, portable heaters, electric blankets, and fans.

Step 3: Buy a Voice Assistant

This is the part that will make your home feel smart. The ability to control your smart bulb and smart plug by voice control will make you feel more in control. Get a Google Assistant–powered speaker like a Google Home, or an Alexa-powered speaker like the Echo Dot. Or, if Apple is your thing, get a HomePod. If you are having trouble deciding between the Google Assistant and Alexa, I recommend Google Assistant for most people. But you really can't go too wrong with either, considering how inexpensive they are.

Step 4: Putting It All Together

Now that you have a smart bulb, a smart plug, and a voice assistant, it is time to put them all together. Make sure you link your smart bulb and smart plug with your voice assistant so you can control them by voice. Play around with the apps that accompany the smart devices to set up automations.

Once you have all of these devices up and running, you truly have a smart home. You can have lights dim or change colors at night. You can have your coffee machine start automatically early in the morning. You can set up night time and morning routines that combine news, weather, and lighting. With just these few devices, you already have a ton of possibilities.

Growing Your Smart Home

Now that you've got your smart home ready, you are probably starting to see the possibilities and have priorities for what you want to do in your home. If you are still looking for inspiration here are a few ideas:

- House and room temperature control with smart thermostats and fans
- Locking doors and arming alarm systems
- Remote notifications from smoke and carbon monoxide detectors
- Controlling irrigation systems
- Turning on security lights
- Making sure garage doors are closed
- Controlling your audio/video system
- Controlling room and seasonal lighting

The next big step to take in growing your smart home is moving from remote control to automation. Most people focus on remote control—the ability to use apps and voice control to command things in your house to happen. However, the true goal of the smart home should be automation, where your home performs action automatically based on things that happen in your home, not commands you have to give it every time. This is where the "smart" in smart home really comes in.

Pick Your Protocol(s) for Growth

The biggest problem with growing your smart home, aside from the cost, is picking from the myriad of devices in the market that accomplish the same thing. Go on Amazon, search for a smart bulb, and you will find a dizzying array of results that operate on different protocols and are compatible with different devices. Some are more open and can work with Alexa and Google Assistant and many smart home hubs. Others work with their proprietary app only. Many will work over Wi-Fi, but you'll also find Bluetooth, Zigbee, Z-Wave, Insteon, and many other protocols. How do you know which types of devices to buy? First, you need to have a good understanding of the different protocols of smart home devices. Here is a listing of the popular home automation protocols of today.

Protocol	Description	Pros	Cons
Z-Wave	Uses low radio frequency band to create a mesh network for multi-device communication	Not very susceptible to interference Reliable mesh network Compatible devices from multiple manufacturers	Occasionally, slow-to-respond network Usually requires a separate device or hub to implement
Zigbee	Uses a 2.4 GHz radio frequency band to create a mesh network for multi-device communication	Reliable mesh network Compatible devices from multiple manufacturers	Occasionally, slow-to-respond network Usually requires a separate device or hub to implement
Wi-Fi (IP)	Radio frequency protocol for connecting fixed and mobile devices to IP networks	Very widely used and understood	Requires a decent amount of power (a problem for battery-powered devices) Susceptible to interference

Bluetooth	Wireless technology standard used for exchanging data between fixed and mobile devices over short distances	Easy to set up Good security Familiar to most people Two-way communication	Limited range
Insteon	A wired or wireless protocol that uses a home's electrical wiring and radio frequency to create a dual mesh network	Good reliability allowed by dual mesh	Not widely available products Professional installation required for some devices Compatible hub required
Infrared	Infrared light used to transmit codes to an infrared receiver	Reliable technology used by a lot of traditional devices like TVs, stereos, and universal remotes	Typically, one-way communication Requires line of sight for transmission Short range

Radio frequency	Generic radio frequency communication on various bands (often 433 MHz)	Low-cost devices	Many different standards and frequencies Offers only one-way communication Requires a radio receiver compatible with sending device's frequency

These are the most popular protocols. The tough part is deciding which one to pick. In my opinion, the three protocols you should focus on are Wi-Fi, Z-Wave, and Zigbee. Most devices today are Wi-Fi because they are well understood. Most smart switches, smart plugs, smart light bulbs, and smart thermostats are Wi-Fi devices. However, Wi-Fi devices don't work well without a constant power source because they require a lot of power. You won't find too many battery-powered Wi-Fi devices. Sometimes, you need to put devices like motion and temperature sensors in places where you don't have power.

The other problem with Wi-Fi is that there is no guarantee of compatibility with your other devices. Wi-Fi specifies how the device connects to a network, but not how it will communicate with other devices. A lot of Wi-Fi devices aren't interoperable with other brands.

Z-Wave and Zigbee solve battery and compatibility problems. Their radios don't require a lot of power to run, making them ideal protocols for battery-powered devices. All Z-Wave devices are designed to be interoperable, as are Zigbee devices. Their primary drawback is that you need some type of compatible smart home hub to control them. We'll talk about smart home hubs just a bit later in this book.

I prefer Z-Wave to Zigbee. There are more Z-Wave devices, and Zigbee is more susceptible to interference, sometimes making their devices less reliable. But I think most users who are looking to build a robust smart home should choose a smart home hub that can work with Z-Wave, Zigbee, and Wi-Fi devices.

What I Do

Over the years, I've ended up with a variety of protocols. I started out by building an extensive Z-Wave network of light switches and smart plugs. My smart door locks are also Z-Wave. Over time, I've added a few Zigbee devices and many IP (mostly Wi-Fi) devices. I also have some radio frequency devices. These consist mostly of cheap push buttons that send a signal telling my smart home to do something. I stay away from IP devices that require a proprietary app, because the ability to integrate devices using a smart home hub and control them locally over my network without relying on a cloud Internet connection is my primary criterion for selecting a smart device. I have found my Z-Wave and IP devices to be very responsive and reliable. I have had occasional problems with my Zigbee devices, probably due to interference.

Now, when I buy a new device, I tend to look for IP (Wi-Fi) devices that can be controlled locally, because my network is solid, and I prefer to be able to integrate all devices with my smart home hub. Sometimes, I can't find an IP device that fits this criterion, and that's when I end up with a Z-Wave or Zigbee device.

Pick Your Smart Home Hub

I've mentioned smart home hubs a few times in this book, but this is the section where I go into detail about what they are and what they can do for you. To recap, a smart home hub integrates and facilitates the automation of the various smart home devices in your home. Many homes have a collage of smart home devices that aren't aware of each other. For example, it's common for a smart home to have a combination of smart lights, smart switches, smart speakers, thermostats, and various other smart devices running on different protocols (e.g., Zigbee, Z-Wave, IP/Wi-Fi, Infrared, RF, Bluetooth—as detailed in the previous section). You may have a separate app to control each of these devices. But what happens when you want one device to trigger an action from another? Or you don't want to have to switch through multiple apps just to do basic things? This is where a smart home hub comes into play.

Different Types of Smart Home Hubs

There are many different types of smart home hubs and almost as many ways to categorize them. I'm going to put them into four categories: hardware hubs, software hubs, smart speaker hubs, and combo hubs.

Hardware hubs

Hardware hubs are the more traditional hubs. They are usually small devices that connect to your home network and interface with all of your smart devices. Each has its own user interface and ways of customizing and configuring. They come with radios (or have add-ons you can purchase) to interface with popular home automation protocols like Zigbee and Z-Wave.

Advantages

- Hardware hubs are easy to install (plug them in and connect).
- They usually have decent-to-good documentation.
- Hardware hub device support out of the box is usually good.
- They offer expansion add-ons for additional device support or capabilities.

Disadvantages

- Hardware hubs often require Internet access to work properly and send your commands to the cloud. (There are exceptions like Hubitat.)
- They can be complicated to configure.
- The backing company controls features and access and can decide to stop supporting it or even stop it from working at any time (e.g., Iris shutdown by Lowe's).

Hardware hub examples

- Vera
- Samsung SmartThings
- Wink
- Hubitat
- Software hubs

Unlike hardware hubs, software hubs separate the software needed to run your smart home from the hardware. Software hubs require you to install the software onto hardware (like a PC or Raspberry Pi), so you need to have some familiarity with installing and configuring programs. In order to interface your software smart hub with Zigbee and Z-Wave components, you need extra hardware to add those capabilities to your computer. These are usually inexpensive Z-Wave and Zigbee USB sticks that you can just plug right into a USB slot on your computer.

Many software smart hubs are free and allow you to adjust your smart home and automations to a degree finer than that of hardware hubs. However, this control also comes at the price of a much steeper learning curve.

Advantages

- There are many fully featured free software smart hubs.
- You can select your own hardware.
- Software systems allow for greater flexibility in customization (user interface and automation).
- Open-source systems usually have great community support.

- You can add missing or additional features yourself if you have the technical chops.
- Software systems are more likely to keep your smart home control local.
- Many systems can interface with hardware smart hubs.

Disadvantages

- Often, there is a steep learning curve.
- The software must be installed on an always-on computer.
- You need additional hardware to communicate with some smart devices.
- You are responsible for keeping the software upgraded.

Software Hub Examples

- Home Assistant
- openHab
- Domoticz
- HomeSeer (can also be purchased as a hardware hub)

Smart Speaker Hubs

Smart speaker hubs are much newer than software and hardware hubs. Many people don't even think of their Google Home or Amazon Echo as a smart home hub, but they are. As these systems have grown in integrations over the years, they have begun to take the place of traditional smart home hubs. Smart speakers can now interface with home security cameras, thermostats, smart locks, lights, and more.

They have also added automations (routines) that allow for triggering multiple actions. Amazon even makes a smart speaker that can directly integrate with Zigbee devices (the Echo Plus).

If you are going to use a smart speaker as your hub or just to control some devices, you need to be thoughtful about how you name your devices. Voice recognition is improving every day, but you'll still have problems if you name devices similarly.

Advantages

- Voice control is included out of the box.
- You already have one and are used to using it.
- Setup and control are easy.
- They seamlessly interface with streaming devices like smart TVs, FireTV devices, and Chromecasts.

Disadvantages

- Limited customization of the user interface and automation.
- Requirement of additional hardware to control Z-Wave and Zigbee devices.
- Processing of commands in the cloud before controlling your devices.
- No Internet connection, no smart home control.

Smart Speaker Hub Examples

- Alexa: Amazon Echo, Echo Dot, Echo Plus, Echo Show
- Google Assistant: Google Home, Nest Home Mini, Nest Home Hub, Lenovo Smart Display
- Siri: HomePod

Combo Hubs

Like smart speaker hubs, combo hubs have a primary purpose in addition to being your smart home hubs. Smart speaker hubs are combo hubs, but they are prominent and different enough to have their own category. Aside from smart speakers, the most common combo hubs are routers and home security systems that are also smart home hubs.

Advantages

- One less vendor/new system to deal with
- Familiarity with a particular system that can double as a hub, making it easier to configure

Disadvantages

- Combo hubs are usually limited in their smart home customization abilities.
- You are locked into one vendor for two pieces of functionality. What do you do in the future when you want to upgrade only one part?
- Combo hubs are more expensive, although they often cost less than two separate systems.

Combo Hub Examples

- Samsung SmartThings Wi-Fi (hub plus mesh Wi-Fi router)
- Samsung SmartThings ADT Kits (hub plus security system)
- Harmony Hub (hub + smart remote) (Add the extender for even more smart home hub capabilities.)
- Almond 3 (hub plus mesh Wi-Fi)
- 2Gig Go!Control (hub plus security system)

How Should I Pick a Smart Home Hub?

When going through the process of selecting a smart home hub, you should ask yourself the following questions:

- Will it work if the company goes out of business?
- Does it keep as much information as possible on my local network and not on the Internet?
- Does it support Zigbee, Z-Wave, and IP protocols?
- Is there active development and an active community for support?
- Is the user interface easy to use and customize?
- How hard is it to configure?

When selecting a smart home hub, think about how much time you have, what kinds of automations you will do, and what smart devices you plan on purchasing. Here are my general recommendations:

- For most people, I recommend the Samsung SmartThings hub. It is popular, straightforward, and compatible with many devices.
- I recommend a Google Assistant– or Alexa-powered smart speaker for those who want simple routines and great voice control.
- Home Assistant is best for those who want their systems to work without Internet access and want to keep control on their home network. You have to be prepared for the learning curve, but you'll have ultimate customization capabilities. Consider Hubitat if you want a similar solution in hardware hub form.

Smart home hubs truly are the key to unlocking the potential of your smart home. They automate and integrate and allow you to customize your home to your needs. Hardware hubs can get you up and running quickly, while software hubs can keep you up all night customizing to your heart's content. If you have only a little time to spend on your smart home, maybe a smart speaker is all you need.

What I Do

I started off with the Veralite hardware hub. It's no longer being made, but you can buy newer versions of it. It worked out pretty well for me, but over time I found its functionality limiting, and eventually Vera discontinued support for it. I decided to look for a new smart home hub and stumbled upon the software hub Home Assistant, and I couldn't be happier with it. It's the perfect smart home hub for the following reasons:

- **It allows for local control.** Most smart home hubs are reliant on cloud servers to function. What happens when those servers are down? What happens when your Internet connection isn't working? Your home goes from smart to dumb.
- **It has tons of integrations**. It seems Home Assistant integrates with everything, and if it doesn't, just wait for a release or two and it will.
- **It has a great user and developer community**. One of the reasons I first started using Home Assistant was because I received quick responses and support to my questions about it when I was searching for a new hub. The communities in the forums and on Twitter are extremely helpful. There is an informative podcast published with each release, which explains the ins and outs of new features and changes.
- **I pick the hardware**. Home Assistant, unlike mainstream hardware hubs, runs on the system of my choice—Linux. It is a Python program. It also runs as a Docker container. Because of this, it can run on a variety of hardware and operating systems, including Raspberry Pis, QNAP NASs, Macs, Windows, and even in virtual machines. You can easily use hardware you already have or purchase an inexpensive Raspberry Pi or NUC to get yourself going.
- **An easy path to voice control**. Integrating Home Assistant with Google Assistant and Alexa can become pretty complex. Doing it yourself requires you to expose your Home Assistant instance to the Internet, which requires proper security controls. If you don't want to set this on your own, you can safely and securely integrate Home Assistant with Google

Assistant and Alexa with a few points and clicks by signing up for Nabu Casa's $5/month service.
- **Powerful automation capabilities**. You can integrate and automate all of those smart devices in any way you can dream up the logic.
- **It's free!** Not sure what more I need to say here. You get an amazing product that costs you only your time and imagination. Whereas most other hubs will cost you money and will take no less time and imagination to install and configure.

Purchasing Future Equipment

Congratulations! Your smart home is set up and things are working great. Like everything related to technology, new and improved devices are always coming out. How do you know what to buy?

The most important consideration is to make sure every new device is compatible with the smart home you have or the smart home you want to have if you are looking to upgrade. This primarily means making sure your device can communicate with the smart home hub of your choice. If your smart home hub supports Z-Wave and Zigbee, get only those kinds of devices. If you are using a voice assistant as a smart hub, make sure you buy devices that are compatible with your voice assistant of choice.

It's hard to predict the future. You don't really know what your smart hub of the future will be. The best you can do is buy devices that support protocols you have been successful with. Also, whenever you are buying an electronic device, try to find a version that can connect to your network. Even better, one that you can control locally.

Locally-controlled devices will probably have some way of connecting to a smart hub. No technology is future proof. However, if you follow these tips, you'll have a much easier time maintaining your smart home over years of smart device purchases. The smart home future is limitless, and you are now prepared for it!

Building a New Home?

About eight years ago, we moved into our current smart home. It was a new construction, but the build was already in progress when we purchased the home. However, I did have an opportunity to specify a few technology things like Ethernet jack, speaker wire, and some outlet locations. Unfortunately, I had only a day's notice and didn't have enough time to really plan right. Also, I had only dabbled in smart home tech before this home. All in all, I think I did a decent job. But as I've progressed in my smart home evolution, I've learned so much more.

There are so many things to think about when building a new home. What colors should the house be? What material should the floors be? How should we lay out the kitchen and bathrooms? These take a lot of time to research and decide. You should also put a lot of time and thought into the smart home capabilities of your home. You can make your smart home build so much easier if you take a few steps during construction. Here are some things to think about if you are fortunate enough to build a new construction smart home now.

Networking New Construction Smart Home Tips

Laying networking wire is least expensive during construction. Here's what you should do:

- Try two Cat 6 or Cat 7 runs to every room. In this day and age, even the bathrooms should be cabled (you never know). It also may be a good idea to run Cat 6 to a central location on the ceiling or high on a wall on each floor for an access point like the TP-Link EAP 225. If you have a larger house or lots of wireless devices, you may want to consider two locations on a floor.
- Consider runs in locations where you may want to install video surveillance cameras. This includes outdoor cameras. This will allow you to use Power over Ethernet (POE) cameras.
- Think about where you will install smart TVs, projectors, and streaming boxes. Make sure you have Ethernet connections in these locations (and power). These may be in the middle of a wall for a mounted TV, or up on a ceiling for a projector. Sure, all these devices could connect via Wi-Fi. But any place you can use wired you should. Try to reserve Wi-Fi for truly mobile devices (cell phones, tablets, laptops, etc.) as much as you can.
- Make sure potential ISPs make connections or create an easy way to make a connection to your home. At my house, cable (coax) was connected from the street to my house, but fiber was not. However, they did install a couple of underground tubes that ran to

my house from the utility pole, which allowed me to install fiber Internet.

I'll talk more about where all these network cables should terminate in the next section.

A/V and Electrical Infrastructure New Construction Smart Home Tips

I have a small structured wiring panel inside my master closet. All of the Cat 5/Cat 6 and coax throughout my house terminate here. It's a small and inconvenient spot behind some clothes. When it was first installed, there wasn't even power, making things difficult. If I could have decided how that went, I'd design an entire closet (maybe around the size of a linen closet) that had power and could fit a server rack. This would make working with the house networking so much easier. Here are other infrastructure items you should consider:

A/V Infrastructure

- Run conduits in the walls of rooms where you have A/V equipment. Sure, HDMI is the standard for A/V connections now. Not too long ago, it was component with coax/optical audio. What will it be in the future? If you run a conduit between where your A/V receiver and display devices will be located, you'll be ready for whatever the future holds, without having to get inside your walls again.
- Run speaker wire in rooms where you'll want surround sound. Or where you think you may want it

someday. Consider running some outside for outdoor speakers.

Electrical Infrastructure

I have spent as much, if not more, time working with electrical systems in my house as I have programming my smart home. Here's a long list of things to consider:

- Think about where outlets should go. Are you going to mount a TV on a wall anywhere? Put an outlet in the wall there so the installation can be clean (along with your conduit for A/V and/or network cables). A projector in the ceiling? Same thing. How about Christmas lights on the outside of your house? Having an outlet near your roof on the outside may not be a bad idea. LED lighting around cabinets and ceilings? What about a smart toilet?
- Wire your home for a generator and/or whole-home UPS. You can prepare your electrical panel for this ahead of time.
- Install a smart panel and/or a smart whole-home energy monitor. While you are at it, install a whole-home surge protector to protect your home's electrical equipment.
- Install large gang boxes in light switch and outlet locations. Smart devices often require more room in the box than regular switches do.
- It's probably a common electrical code everywhere now, but make sure your switches and outlets have a neutral wire. Most smart switches require neutral connections.

- Make sure all thermostat locations have a C-wire. Most smart thermostats (e.g., ecobee) need a C-wire to work properly. You'll be able to easily install a smart thermostat if you install a C-wire at the thermostat location in advance. While you are at it, make sure you plan enough zones to ensure even heating and cooling in your home.

Other Infrastructure

- Consider a smart water monitoring/shutoff valve.
- Do you want motorized blinds? Consider running power to blind installation locations.
- If you want wired security sensors (door sensors, motion detection, window sensors), have these installed at build too. You won't have to worry about batteries.
- Interested in an electric or hybrid vehicle in the future? Install a charger.
- Consider installing wiring for landscape lighting.
- If you are interested in an irrigation system, look into a smart irrigation controller, and make sure you have good power and network access where you want to install it.

Even though building a new construction home is exciting, it takes a lot of work and decision making. I suspect someday many of these suggestions will be standard for new house development. The smart home of today is the regular home of tomorrow.

Key Takeaways

There are many different types of smart devices that you can use in your smart home.

- ☐ You should inventory the smart devices you already have as you plan your smart home build.
- ☐ Test the waters with a few smart devices before doing an extensive smart home build.
- ☐ Make sure you understand the pros and cons of the different smart device protocols.
- ☐ The smart home hub is the brains of your smart home. Take care in choosing the right one that will support all your smart devices and needs.
- ☐ Make sure future smart devices you purchase are compatible with your smart home hub.
- ☐ Build smart home features at the start if you get the chance to build a new construction home.

A Day in the Life: Good Evening in Your Smart Home

5:30 p.m. – You pull into your driveway. Your home recognizes that it's your car and automatically opens the garage door for you. As you walk in the house the kids ignore you, but your smart speaker welcomes you home with a cheery message.

6:00 p.m. – You decide it's time to relax in your garden and listen to music. You're glad your outdoor wireless access point gives you a strong enough signal to stream music in your garden.

6:40 p.m. – You get an alert on your smartwatch that someone is at the door. You check the video doorbell app on your phone. The pizza has arrived! Time to go eat.

7:15 p.m. – After dinner, it's time to watch your favorite family TV show. "Hey Google, turn on Netflix," you say to your smart speaker. Lights automatically dim around the room as the TV and A/V receiver turn on and automatically put themselves on the right inputs and proper volume for Netflix.

7:30 p.m. – Sunset has arrived and the outdoor security lights automatically turn on.

9:00 p.m. – The lights blink in the family room, indicating it's time for the kids to go to bed. The Wi-Fi connection to the kids' smartphone and tablet automatically disconnects for the night.

9:30 p.m. – After having a glass of wine, it's time for you to head upstairs for the evening. You say, "Ok Google, good night." In response, the TV and A/V receiver turn off, the alarm arms, the doors lock, garage doors and blinds close, and a path of lights to your bedroom turns on.

10:00 p.m. – About thirty minutes after your good night command, all non-security lights turn off. The thermostat saves energy and no longer regulates your temperature for the night. Time to rest up for tomorrow.

Home Automation Ideas and Inspiration

The things you can do with a smart home are limitless. But since the sky's the limit, it can be difficult to know where to start. Here are a few ideas of things to automate in your home to help you get started.

Night/Security Lights

There is a lot of talk about alarm systems, but some of the best security you can have is simply keeping the area around your home well lit at night. Make this happen every night from dusk until dawn with a home automation that turns your outdoor lights on and off. That's easy enough to do with most smart bulbs on the market, or you can install smart switches for your outdoor light bulbs. I still use timer switches for some of my outdoor lights (if it ain't broke...).

Vacation Lighting

Take the security lights a step further by having lights come on during vacation. You don't want your house to look abandoned for a long period, as that is an invitation for thieves. Use smart bulbs and smart switches to turn your lights on at strategic times of the day when your house is empty to make it look like someone is home.

Turn off Lights When Arming Alarm "Away"

Home Automation can enhance your alarm system quite a bit. Most alarms have "home" (someone is still in the house) and "away" (no one is in the house) arming modes. The difference is usually that motion detectors aren't enabled when you arm "home" (because you aren't perfectly still when you're home!) and are enabled when you are away. But how about your lights? My kids often leave lights on when we all leave the house, but we never forget to arm the alarm. How about an automation that turns off all the indoor lights when you arm the alarm "away"? You could also program it to leave a few on for security.

Lock the Doors When the Alarm Is Armed

Locks are another popular home automation. When you arm the alarm ("away" or "home"), you aren't expecting anyone who doesn't live in the home to come in through a door without disarming the alarm first. So why not trigger locking the doors whenever the alarm is armed (just in case you forget)? You'll need a smart lock for that. I use an older, Z-Wave–enabled lock to do this, but you have a lot of options.

Turn on Lights When the Alarm Is Set off

You most likely have a siren that goes off when the alarm is tripped, but you can enhance this by turning on strategic lights when your alarm is set off. This might help scare off burglars and make sure you can see what is going on.

Send a Text When the Alarm Is Set off

If you have third-party alarm monitoring, you probably have this feature. But if you don't, it may be good to make sure you also get a text or some other notification when your alarm is set off.

Automate Christmas Lights

Using timers to automate Christmas lights is a strategy that has been around for decades. These days, I plug all my Christmas lights into smart switches and have Home Assistant control when my lights turn on and off. My outdoor house lights are color-changing smart bulbs that I can control from Home Assistant. I have them change to Christmas colors when my Christmas lights come on. This is one of my favorite home automations.

Notification When Someone Arrives Home

Want to know when someone arrives at your house? Set up presence detection and have it text you. This is great for checking on kids or on loved ones you are worried about. There are many ways to do presence-detection, including Bluetooth scanning, network scanning, and third-party services like OwnTracks. I just scan for when phones are on our home network.

Motion-Activated Lights

It's just too much work to walk all the way to inconvenient light switches or yell out voice commands to Alexa or Google Assistant. Why not just have lights turn on when they detect you? A motion detector and a smart switch or bulb (or a motion-detecting light switch) are all you need to get this done. Great for hallways!

Coffee Ready in the Morning

Are you a creature of habit? Are you a monster before your coffee? You can have your coffee maker brew a hot pot at a certain time of the morning each day. You can do this in a lot of ways. Some coffee makers start brewing when they turn on. In this case, just use a smart switch or a timer to have it turn on at a certain time. Other coffee makers have timers and schedules built-in.

Lock Doors Automatically

I know I mentioned locking the doors when the alarm is armed. But what if you don't have an alarm? You can set your smart locks to lock at a certain time each night.

Wake up to Soft Lights, Music, and the News

The sound from many alarm clocks is so jarring! How about you ease into your day by automating your alarm to slowly brighten the lights, turn on some music, and then give you a news briefing? One easy way to do this is to set up a good morning routine with Google Assistant.

Vacuum the House When No One's Home

Robot vacuums have been around for a while and continue to get better and less expensive. Many robot vacuums allow you to schedule them to clean at times when you aren't home.

Automate the Temperature of Your House

This one can save you money. Have your thermostat (1) lower the temperature when you are sleeping or away from home and (2) raise it when you are awake in the home. Some thermostats (e.g., Google's Nest thermostats) can even learn your routines and create this automation for you.

Reboot Devices and Computer Systems When They Are Unresponsive

If you have devices that you leave on all the time, but every so often they freeze, this may be a home automation for you. If it is a computer, you can implement this using wake-on-LAN or a setting in the BIOS to turn on the computer in the event of power loss and restoration. Then you can turn the computer on with the toggle of a smart switch. You can do the same with other devices. The key is having a reliable method for determining when the device isn't functioning.

Notification When Your House Is Using Too Much Electricity

There are many devices that can monitor and report your home electricity usage. You can set these up to signal when you've passed a daily or monthly electricity threshold so you can watch your usage more closely.

Open Your Garage Door When You Drive Up

This one is only for those who are confident about their ability to automate. You can set your garage door to automatically open when your presence detection (using Bluetooth, third-party apps, network detection, or other) detects that you and your car have just arrived home. There are many ways to incorrectly trigger your door opening, so you want to make sure your logic is correct. There are a lot of smart garage door openers. I opted for a DIY route by wiring in a Shelly 1 relay to my garage door opener and triggering it with my home automation controller.

Turn on Indoor Lighting at Sunset

Controlling lighting is an extremely popular home automation task, because there are many simple things one can do. For instance, you can have particular lights inside your house automatically turn on as it gets dark at sunset. Then you don't even have to get up to turn the lights on while you are watching TV.

Color Indicators to Tell Early-Rising Kids to Go Back to Sleep

Here's one that was a lifesaver when my kids were really young. Young kids like to wake you up when they wake up at night, no matter the time. We installed a clock that changes color to green when it is okay to wake us up. You can do this with a clock or with a smart light in your child's room.

WHAT NEXT?

KEEP ON BUILDING. KEEP LEARNING about new smart home technologies and keep tinkering with your smart home devices and automations. Stay up to date on new technologies that are coming out every day. Look for chances to improve your smart home. Most importantly, have fun and enjoy the fruits of your labor!

Appendix A

Example of Smart Home Buildouts

Quick and Inexpensive Starter Smart Home

This buildout is just to get your feet wet in smart home technologies and capabilities. If you aren't sure you want to do one, this is a good place to start to give you some rudimentary smart home capabilities. You can have a functioning smart home with one of each of these devices for less than $100.

Device	**Purpose/ Capability**	**Example Devices**
Voice assistant	To enable voice control of devices and to control multiple devices with one command.	Google Home Mini Amazon Echo Dot
Smart light bulb	To add lighting effects to a room. Allow scheduling and remote control of the lighting.	Kasa (TP-Link) Bulbs LIFX Bulbs Philips Hue Bulbs (requires hub)
Smart plug	To enable remote and scheduling ability to an outlet (coffee maker, lamp, Christmas lights, heater, etc.).	Kasa (TP-Link) smart plugs Wemo smart plugs

Smart Home Safety and Security Buildout

These are some items you might want to consider if you want to add security smart home features to your house. This includes video surveillance, automated locks, and more. Building out a full-blown security system with all of these devices could easily run you $1,000, but if you bargain hunt and build over time, the costs can be kept much lower.

Device	Purpose/Capability	Example Devices/Brands
Surveillance cameras	To monitor activity going on inside and outside of your home.	Wyze Cam Ring/Nest Video Doorbell
Network Video Recorder (NVR)	To record video from your surveillance cameras. It can be hardware or software-based.	Blue Iris Zoneminder Ubiquiti UniFi NVRs
Smart deadbolts	Allow for remote and automated locking and unlocking of exterior doors in your home.	Yale Kwikset Schlage
Smart door sensors	These devices can sense when a door or window is open or closed in your home. This can be helpful in detecting intrusions or knowing whether you've left a door open that you didn't mean to.	Samsung SmartThings Aeotec

Smart switches/ Smart bulbs	These can be used to (1) automate outdoor lights to turn on when it is dark and (2) using indoor light automation, to mimic that you are home when you are away for vacation.	Kasa (TP-Link) Wemo LIFX
Flood sensors	These can notify you when there is water where there shouldn't be. For instance, I keep a flood sensor in my basement.	Aeotec Fibaro FortrezZ Ecolink
Smart security system (door sensors, motion detectors, glass break sensors, siren, etc.)	These are enhanced security systems that can also tie into and/or control smart systems in your home. They can also be professionally monitored. Many come with options for other security items in this list (e.g., door and flood sensors).	Ring Alarm SimpliSafe Alarm.com
Smart home hub	You're definitely going to want a smart home hub to tie together and automate multiple devices from multiple manufacturers.	Home Assistant Samsung SmartThings Hubitat Vera

Lighting Buildout (Including Christmas Lights)

Wouldn't it be great if your lights came on when you wanted them to, turned off when you wanted them to, and you didn't have to intervene? Depending on the regularity of the schedules of the people in your home, it may not be feasible to totally automate your lights, but I guarantee you can come close. A few hundred dollars can get you far!

Device	Purpose/Capability	Example Devices/Brands
Smart switches /Smart bulbs	These can be used to automate outdoor lights turning on when it is dark, as well as using indoor light automation to mimic that you are home when you are away for vacation. You can also use smart bulbs to automate the colors and brightness in a room for different times of the day	Kasa (TP-Link) Wemo LIFX

Motion sensors	These can be used to detect someone's presence (or lack thereof) in a room and then turn the lights on or off as appropriate.	
Smart plug	To enable remote and scheduling ability for things like Christmas lights and lamps.	Kasa (TP-Link) smart plugs Wemo smart plugs
Smart Christmas lights	These days you can buy Christmas lights that are smart (or if you are more adventurous, you can make them on your own)	Twinkly Philips Hue Altove LED lights NodeMCU (ESP8266)
Smart home hub	You're definitely going to want a smart home hub to tie together and automate multiple devices from multiple manufacturers	Home Assistant Samsung SmartThings Hubitat Vera

Heating and Cooling System Buildout (temperature sensors, humidity, smart thermostat, fireplace)

One automation category that can actually save you money over time is your heating and cooling system. If you have a complex setup, you can also use a smart home hub to better coordinate and control your HVAC system. For instance, you could automate your gas fireplace to come on for localized heat instead of your HVAC. Unfortunately, smart thermostats aren't cheap. They can easily cost you $150–$300 each.

Device	Purpose/Capability	Example Devices/Brands
Smart thermostat	Add smart scheduling and remote control (at "home" and "away") to your HVAC system.	ecobee Nest Honeywell
Temperature and humidity sensors	Oftentimes thermostats aren't installed in the best places for detecting temperature. A remote sensor can solve that problem. Many smart thermo-stats have remote sensors made to work with the thermostat. These are often bundled together.	ecobee Nest Aeotec Fibaro Samsung SmartThings

Smart fire-place control	These allow you to automate and remotely control your fireplace. This is an overlooked part of your HVAC system.	Durablow WiFire A smart dry contact relay
Smart HVAC system	Some HVAC systems have smarts built right into them. They connect to the cloud and allow you to control them with an app without the need for a special thermostat. Look for these options if/when you are overhauling a part of your HVAC.	Honeywell Mitsubishi Many others

Appendix B

Checklists

First-Year Tasks and Goals

Months 1–2:

- ☐ Learn about smart home and smart home capabilities.
- ☐ Learn about voice assistant and smart speaker capabilities.
- ☐ Think about tasks around the house you'd like to automate.
- ☐ Inventory your current smart home devices.
- ☐ Make sure your home network and Wi-Fi are strong, stable, and secure.
- ☐ Learn about popular smart home protocols (Zigbee, Z-Wave, Wi-Fi, etc.).
- ☐ Buy, install, and configure a smart bulb, smart plug, and/or smart switch.
- ☐ Buy and configure a smart speaker (Nest Home, Amazon Echo, Apple HomePod).
- ☐ Configure an automation or two (e.g., automatically turning lights on and off).
- ☐ Set a first-year smart home budget.
- ☐ Months 3–4:
- ☐ Learn about smart home hubs.
- ☐ Select a smart home hub that supports the smart home protocols you want to use.
- ☐ Buy more smart home devices (motion detectors, smart thermostat, smart door lock, smart switches, smart bulbs, etc.).
- ☐ Think of something really cool that you'd like to install and automate around the house (e.g., LED lights, smart garage door).

- ☐ Configure a few automations across multiple smart devices (e.g., automatically turn lights on in a room when motion is sensed).
- ☐ Think about your medium – and long-term smart home goals.

Months 5–8:

- ☐ Continue to automate tasks around the house.
- ☐ Talk to household members about what tasks they'd like to automate.
- ☐ Enjoy what you've automated so far!
- ☐ Keep up to date on new and evolving smart home technologies and capabilities (use some of the resources in Appendix D).

Months 9–12:

- ☐ Evaluate your smart home hub. Is it capable of integrating and automating all of your smart devices?
- ☐ Evaluate how well your smart home has worked. What would you improve? What do you really like?
- ☐ Revise your long-term smart home vision.
- ☐ Create your medium – and long-term smart home implementation plan (goals, budget, smart devices).

Current Smart Home Device Inventory Checklist

Check to see whether you already own the following smart home devices. They may give you a jump start in building and automating your smart home.

- ☐ Smart speaker
- ☐ Timer switches for lights or fans
- ☐ Motion detectors
- ☐ Security system
- ☐ Smart bulbs
- ☐ Smart plug
- ☐ Smart TV
- ☐ Streaming box (Roku, Fire TV, Android TV Box, Chromecast, etc.)
- ☐ Video game console (Xbox, PlayStation, Nintendo)
- ☐ Raspberry Pi
- ☐ Irrigation system and/or irrigation timers
- ☐ Robot vacuum
- ☐ Network-connected A/V (audio/video) receiver
- ☐ Network-connected printers
- ☐ Other network-connected devices________________________

Wi-Fi Router Checklist

- ☐ Wi-Fi router features to look for:
- ☐ Support for wireless AC
- ☐ Supports MU-MIMO or SU-MIMO
- ☐ Advanced firewall
- ☐ Quality of Service (QoS)

- ☐ Supports WPA2 AES security for Wi-Fi or newer
- ☐ Gigabit Ethernet ports
- ☐ Guest network support
- ☐ Receipt of consistent firmware updates
- ☐ *Not* an ISP-provided model
- ☐ Well reviewed on Amazon or in professional online reviews

Appendix C

Additional Resources

HomeTechHacker Blog (https://hometechhacker.com/) – This is my personal blog where you can find many helpful articles about building and improving your home network and smart home. On my site, you will also find an up-to-date list of recommended smart home and home network devices (https://hometechhacker.com/top-picks-for-common-household-tech/) and updates to this book.

CNET Smart Home (https://www.cnet.com/smart-home/) – This site has an abundance of information and reviews of home technology products.

The Ambient (https://www.the-ambient.com/) – This site has lots of up-to-date reviews on the latest smart home products.

Digital Trends (https://www.digitaltrends.com/smart-home-reviews/) – Their home technology product reviews are thorough and entertaining.

How-To Geek smart home section (https://www.how-togeek.com/t/smarthome/) – This site is full of tutorials, comparisons, and explanations of smart home technology.

Home Assistant Blog (https://www.home-assistant.io/blog/) – I can't truly express how wonderful the Home Assistant smart hub is. If you're a bit technically inclined and like to tinker, the payoff is more than worth the time.

pfSense Documentation (https://docs.netgate.com/pfsense/en/latest/) – I've been really happy with pfSense software running my router. It's free software with lots of documentation and features that can run on hardware of your choice. You can buy a router with pfSense preinstalled. This documentation will get you started.

The Hook Up YouTube Channel (https://www.youtube.com/channel/UC2gyzKcHbYfqoXA5xbyGXtQ) – Great video tutorials and explanations of all kinds of smart home products and projects. Many of the videos are not for beginners but still worth a look for the education.

GLOSSARY

Automation – Automation is the technology by which a process or procedure is performed with minimal human assistance. Home automation means that things in your home happen without you or another person having to make them happen—e.g., your doors automatically locking at night or a light automatically coming on when you enter a room.

Automations (plural) – Automations is a term that smart home builders use to describe multiple sets of automated instructions.

Bluetooth Low Energy – Bluetooth Low Energy, often referred to as Bluetooth LE or BLE, is a form of Bluetooth designed to significantly lower power consumption. Bluetooth is a wireless communication standard that has been around for decades. BLE takes that standard and lowers the power requirements, allowing smaller battery-powered devices to communicate frequently with other devices via Bluetooth without rapidly depleting their batteries. BLE was integrated into the Bluetooth 4.0 standard. It's great for devices that only periodically need to transmit and/or receive small amounts of data.

Cat 5e, Cat 6, Cat 6a, Cat 7, Cat 8 – These are all different standards of Ethernet cable, which are the backbone of the wired network in your home. The higher the number, the later the Ethernet standard. The later the Ethernet standard, the faster speeds the cable will support over longer distances. If you are laying down new wire at the time of my writing this book, try to get at least Cat 6a, which can support a 10-Gigabit connection for 100 meters.

HVAC – HVAC (i.e., heat, ventilation, and air conditioning) refers to the environmental systems in a home that control its temperature and air quality.

Insteon – Insteon is a family of X10-compatible smart home devices that use both radio frequency and power to transmit signals. They are X10 compatible in that they can accept X10 commands but don't repeat them.

IoT – IoT (i.e., Internet of Things) is a collection of interconnected and interrelated devices that can communicate and transfer data over various networks without human interaction. It most commonly refers to the explosion of devices that are everywhere that connect to the Internet. Examples include Smart TVs, smart speakers, toys, wearables, smart appliances, and smart meters.

ISP – ISP (i.e., Internet Service Provider) is a company that provides Internet as a service to a home or business. These are often cable and other telecommunications companies like Comcast, CenturyLink, and Verizon.

LAN – LAN (i.e., local area network) is a computer network that usually connects computers in a specific geographic location—a home, a school, or an office building.

Mesh Wi-Fi Router – A Mesh Wi-Fi router, sometimes called Whole Home Mesh Wi-Fi, consists of a wireless router and satellite wireless access points that provide wireless access throughout your home. The satellite wireless access points connect to the router wirelessly and don't need a wired network connection to the router. This means they can be flexibly placed at optimal points to provide Wi-Fi throughout a home.

Modem – Modem, which is short for modulator-demodulator, converts signals from one type of device to another type of device. One example is a cable modem, which converts signals from a coaxial cable (analog) to an Ethernet signal (digital) that routers typically use.

Router – A router forwards (or routes) packets of data between different networks. For instance, the router in your home routes packets coming from your ISP (WAN) to your private home network (LAN) and vice versa.

Smart Speaker – A smart speaker is a network-enabled speaker that can be controlled by spoken word and can also stream audio content and responses. They can communicate with other smart devices. Popular examples include the Google Home and the Echo Dot.

Smart Home – A smart home is a home that provides some combination of comfort, energy efficiency, security, lighting, etc., aided by technology that allows these systems to be automated, integrated, and available for remote control.

Smart Home Hub – A smart home hub integrates and facilitates automation of the various smart home devices in your home.

Voice Assistant – A voice assistant is a digital assistant that uses voice recognition and natural language processing to respond to spoken inquiries and commands through a device like a smartphone or a smart speaker.

WAN – WAN (i.e., wide area network) is a computer network that covers a large geographic region. They are similar to LANs, but aren't limited to a single location and are a lot bigger.

Wi-Fi – Wi-Fi is the name of a wireless networking technology that uses radio waves to provide high-speed network connections.

Wireless Access Point – A wireless access point connects to a router, switch, or hub via Ethernet, and projects a wireless signal to a designated area. These are used for extending a wireless network.

X10 – X10 is a smart home protocol that uses power line wiring for communication across smart home devices. It's one of the older protocols and not used a lot by itself anymore. Sometimes, you will find them in hybrid systems

(systems that use multiple protocols), because the older devices still work and don't need to be replaced.

Zigbee – Zigbee is a low-power, low-data-rate, mesh wireless networking standard. In mesh systems, signals can hop from one device to another so that signals can cover large distances. Zigbee uses 915 MHz and 2.4 GHz radio frequencies.

Z-Wave – Z-Wave is also a low-power, low-data-rate, mesh wireless networking standard. Z-Wave uses the 800–900 MHz frequency range and thus is subject to less interference than Zigbee.

ACKNOWLEDGEMENTS

I'D LIKE TO THANK MY wife, Adriana, for supporting me through the endeavor of writing this book. I'd like to thank my family for allowing me to tinker to my heart's content around the house and being willing to try (and ultimately love) the new features of our smart home. I'd also like to thank my family for providing input and inspiration for many of the projects I've completed around the house.

I'd like to thank my Twitter followers, newsletter subscribers, and all the people who publish podcasts, YouTube videos, and blogs that have supported me and helped me build my smart home.

Last, I'd like to thank Graham Southorn for helping turn my raw manuscript into a readable and professional book. Every good book has a good editor.

ABOUT THE AUTHOR

MARLON BUCHANAN HAS WORKED IN the IT field for over twenty-five years as a software developer, a college instructor, and an IT Director. In his free time, he can be found researching new smart home projects, watching and playing sports with his children, and writing articles for his blog, HomeTechHacker.com. He holds a bachelor's degree in computer engineering and master's degrees in software engineering and business administration.

Please sign up for his newsletter on his blog. You can also follow him on these social media channels:

Twitter – Twitter.com/HomeTechHacker (@HomeTechHacker)
Pinterest – Pinterest.com/HomeTechHacker
Facebook – Facebook.com/HomeTechHacker

WHAT DID YOU THINK OF THE SMART HOME MANUAL?

FIRST OF ALL, THANK YOU for purchasing *The Smart Home Manual*. I know you could have picked any number of books to read, but you picked this book and for that I am extremely appreciative.

I hope that it has inspired and helped you to build and improve your smart home. If so, it would be really nice if you could share this book with your friends and family by posting to Twitter, Facebook, and Pinterest.

If you enjoyed this book and found some benefit to reading this, I'd love to hear from you. I hope that you can take the time to post a review on Amazon. Your feedback and support will help me to greatly improve my writing craft for future projects.

I want you, the reader, to know that your review is very important, and very appreciated.

I wish you all the best in your future smart home success!